P9-DXH-665

THE NATURAL GEOGRAPHY OF PLANTS

THE NATURAL GEOGRAPHY OF PLANTS

HENRY A. GLEASON AND

ARTHUR CRONQUIST

COLUMBIA UNIVERSITY PRESS / NEW YORK & LONDON

Henry A. Gleason was on the staff of the New York Botanical Garden from 1919 to 1950, much of the time as Head Curator and Assistant Director. Since the beginning of 1951, he has been Head Curator Emeritus.

Arthur Cronquist is a Curator at the New York Botanical Garden.

Copyright © 1964 Columbia University Press

First printing 1964
Second printing 1965

Library of Congress Catalog Card Number: 64–15448
Printed in the United States of America

QK
101
G57

Preface

In the fall of 1961 Dr. Gleason asked me to complete his manuscript on plant geography, to make such changes as I felt necessary, to assemble the illustrations, and to see the book through publication. I have been happy to do so. Chapters 1 through 21, dealing with the theory of plant geography, are Dr. Gleason's work; only Chapter 11 has undergone substantial modification. Chapter 22, dealing with the actual vegetation of North America north of Mexico, is my own, as is the map of the floristic provinces of the continental United States and Canada reproduced in Chapter 15. First person references in the first 21 chapters are Dr. Gleason's; in the last chapter they are mine.

NHS

Many people have helped me in the completion of this work, by providing either information or photographs. The photographic files of the United States Forest Service, under the care of Lee Prater, have been the most important single source of illustrations. These and other photographs are acknowledged individually in the legends. Dr. Elbert L. Little, Jr., also of the U.S. Forest Service, has kindly made available the latest maps of the distribution of certain forest trees; most of the maps reproduced in Figs. 14.7 and 14.8 were drawn from large-scale maps received from him. My daughter Elizabeth served as the model to show the size of the trees in Fig. 22.34. Pertinent information and comment have been received from E. J. Alexander, Rupert C. Barneby, John Beaman, Janice Beatley, Murray Buell, Pierre Dansereau, Howard Irwin, Askell and Doris Löve, Rogers McVaugh, Elsie Quarterman, and David Rogers. Dr. Quarterman was kind enough to let me read in advance of publication the manuscript prepared by her and Catherine Keever on the forests of upland sites in the coastal plain. To all of these people, I express my appreciation. Responsibility for what is said in the following pages naturally falls on Dr. Gleason and me, and not on those who helped us.

INTERSTATE
SEP 28 '65

ARTHUR CRONQUIST

April, 1964

76859

Contents

Introduction 1
1 What Is Plant Geography? 5
2 Distribution and Range 13
3 The Migration of Plants 27
4 The Principle of Discontinuous Distribution 51
5 Some Considerations of Seed Production 61
6 Behavior of a Plant within Its Habitat 71
7 The Speed of Migration 81
8 Retreating Migration 87
9 Limits of Range and Migration 91
10 Looking Backward on Migration 109
11 Looking Backward Still Further 113
12 Summary of Chapters 2–11; Introduction to the Following Chapters 127
13 The Joint Migration of Species into New Habitats 133
14 The Establishment of Joint Ranges 151
15 Flora, Floristic Group, Province 175
16 The Nature of Floristic Groups and Provinces 179
17 More about Floristic Groups 201
18 Vegetative Form 223

19 The Concept of Vegetation 235

20 Correlation of Climate and Vegetation 239

21 On Transitions between Provinces 263

22 The Vegetation of North America, North of Mexico 275

Index 415

Introduction

Plants and the landscape. Perhaps we would do well to change one word in that phrase and write plants are the landscape. Let us explain this.

We all know that the world is a great ball of rock and metal. It is not a perfect sphere, but its surface is wrinkled into hills and mountains, elevated in some places into great plateaus, depressed in others into broad basins. Over this ball is spread a very thin, superficial layer of soil and water. The water naturally collects in the depressions, and there is so much water that it covers about three-fourths of the surface of the world. Some of it is liquid and forms the oceans and lakes. Some of it is frozen and forms the ice-caps of Greenland and Antarctica. These areas of water, whether liquid or frozen, can not well be called landscape.

The average depth of the oceans is about two miles, but the amount of soil is far less. Very rarely is it more than a hundred feet deep and in many places it is less than ten feet. In all mountainous countries it is interrupted by exposed cliffs and rocks without any soil at all.

This thin layer of soil is everywhere eroded by water, or blown about by the wind, or pushed around by glaciers. Millions of tons of it are washed away by the rivers, permanently removed from the land, and eventually deposited on the bed of the ocean. This loss is compensated by the continual disintegration of the solid rock, which is constantly being worn down by erosion, broken up by frost and heat, ground into powder by glaciers, or decayed by the chemical action of the soil, the air, and the soil-water.

Finally over this thin layer of soil is spread a superficial layer of plants, a thin layer or a thick one, of big plants or of little ones, but in most parts of the world a layer so complete and so continuous that it completely hides the soil.

The whole may be compared to a picture. The solid core of rock, of which the world is composed, is our easel. Over this is stretched a thin canvas of soil, and on the canvas is painted a picture with plants, and this we call the landscape.

When we enjoy a picture, we are concerned chiefly with that thin coating of water-colors or oils. We never think of the canvas, although we recognize that the beauty of the picture may be enhanced by an appropriate frame. When we look at a landscape, we rarely think of the invisible soil, but we admit that the charm of the scene is often increased by the configuration of the land, although the land, without its painting of plants, is bare and desolate. It is the vegetation which makes the landscape.

Like the paints in a picture, plants usually cover the soil completely. If one flies over the country at low altitude, the whole land is green except where fields have recently been plowed. If one walks through the woods in summer, one does not see the soil. It is hidden beneath a carpet of herbs and mosses or covered with the dead leaves of last season. Even in our prairies and plains the plant cover is nearly or quite continuous. Only in our mountains or deserts do we see anything like a real *land*-scape; everywhere else we see only the *plant*-scape. In mountains, the cliffs and rocks are more or less occupied by plants. Many kinds of flowering plants live in the little accumulations of soil in pockets and crevices, while the exposed surfaces are covered, often brightly, with lichens and mosses. In most of our deserts plants are present, though often stunted or sparse, and at certain seasons, after some rain has fallen, the whole desert seems to burst into bloom. In short, nature, the master painter, spreads her colors almost everywhere and combines them attractively to make the Landscape.

One September afternoon we drove around a curve in a New Hampshire road and found ourselves in a traffic jam. A hundred cars had stopped, their occupants were out on the road, and all cameras were in use. Just to the east, glittering in the reflection from the western sun, was a mountainside golden with autumn foliage. As we finally drove away we thought, as botanists might, that probably not more than one out of ten of those spectators had any idea what kind of trees they were looking at. If those same people had driven west from Washington a few weeks later, they would have seen a similar display on the slopes of the Blue Ridge, only in bronze and crimson, and again probably only one out of ten would have wondered why the colors were so different.

It is to this one person out of ten that our book is addressed; to those people who not only enjoy the landscape, but also like to meditate on it, to ask themselves questions about it, to try to understand it, and hopefully to arrive at some conclusions about it.

There are some conclusions about the landscape which are quickly and easily reached by all thinking people. They are doubtless well known already to every reader, but nevertheless we repeat them here, just as a foundation for what is to follow.

The first is that the many different kinds of plants are restricted each to a particular section of the country. As a New Englander drives south for a winter in Florida, he notes with interest the appearance of the first cypress tree, the first live oak, the first tree draped with the somber Spanish moss. As a Kentuckian drives north for a vacation in northern Michigan, he watches for the first white pine or the first tamarack. They (and we) know that the first three of these plants are southern and do not extend to the northern states, while the last two are northern and not found in the south. We conclude, and rightly, that the same holds true for every other kind of tree in the world, that each of them is restricted to a certain part of the world, be that part large or small. We carry our conclusions further and decide, again rightly, that the same principle applies to all smaller plants as well, that the wake robin of New England, the bloodroot of Ohio, the tall rosinweed of Iowa, the bluebonnet of Texas, the columbine of Colorado, the poppy of California are also restricted similarly, each to its own part of the country.

The second of these conclusions has to do with the general form of the plants. The observer can not fail to see, and sooner or later comes to appreciate, that the plant life of some areas looks very different from that of others. The landscape may consist chiefly of trees, as in the eastern and Pacific states, or mostly of grasses, as in the plains states, or mostly of shrubs, as in the intermontane states. He also soon realizes that there may be similar variations even within a limited region, although usually much less conspicuous.

The third conclusion is that the conditions which have been mentioned for the first and second are in some way due to the nature of the surroundings, and one suspects that they are caused mostly by differences in the climate or the soil.

The first two conclusions are obviously true, being based on direct observation. The third conclusion is based on good reasoning only, and is also true as far as it goes. None of the three has been elaborated in sufficient detail to account for all observable facts. It is the duty of this book to explain these matters further, as far as can be done in a limited space and in our present state of knowledge.

Science is merely a term for organized knowledge, although by general custom it is commonly restricted to the knowledge of actual objects and their behavior, as zoology, astronomy, or physics. The particular science to be discussed here is called plant geography, and it is the science which helps us most to understand and appreciate the landscape.

1/ What Is Plant Geography?

"Of all the branches of botany there is none whose elucidation demands so much preparatory study, or so extensive an acquaintance with plants and their affinities, as that of their geographic distribution."

We must not be discouraged by these words, written more than a century ago by Sir Joseph Dalton Hooker, later to become Director of the Royal Botanic Gardens at Kew in England. Hooker was an astute botanist and spoke with authority. After a century of study since Hooker's time, it would seem that a clear understanding of plant geography has become even more difficult to acquire. We know more now, to be sure, but we have uncovered so many questions still unanswered and so many problems still unsolved that we are often amazed at the narrow limits of our knowledge and the great extent of our ignorance. No one today is able to write a really comprehensive treatise on plant geography, although there are those who are willing to try it. An introduction to the subject must really be an introduction.

Plant geography, as the words imply, is the branch of botany which deals with the space relations of plants, particularly in their broader aspects and especially in reference to the distribution of plants over the surface of the earth. That is a simple and easily understood definition. Why then is it not equally simple to learn the salient features of the science?

The concrete facts of plant geography are relatively simple and easily learned, if one is content to take another's word for them and not ask too many questions about them. For a hundred and fifty years, extant books have presented these concrete facts and described the general features of the plant life of the various regions of the world. The gathering of the facts for one's self would naturally require unlimited opportunity for travel into all parts of the world, and into each part during the proper season. Obviously, no individual could observe all the facts for himself.

To learn the general features of plant distribution in the United States by personal observation is, of course, fairly easy in these days of convenient and unrestricted travel. There are thousands of Americans who already know from firsthand experience that Spanish moss (*Tillandsia usneoides*, Fig. 1.1) and live oak grow only in the southern states and the giant sequoias (*Sequoiadendron giganteum*, Fig. 1.2) only in the Sierra Nevada; that Arizona is mostly covered with desert and Iowa with fields of grain. All such matters are the observable facts of plant geography, upon which the science is based. The amount of information which one may easily acquire for himself is limited only by his interest in the landscape and the size of the area which he covers in his travels. But comparatively few people stop to ponder the significance or the causes of what they see and thereby become interested in plant geography as a field of science. It is to these curious or interested persons that this book is addressed.

Thousands of people know, or have known, many of the facts of plant geography, but very few have undertaken to correlate these facts or to analyze the causes of which they are the result. In this book we shall present a small fraction of the known facts, choosing them primarily from the United States, and attempt to explain or suggest some of their causes.

A full understanding of the facts of plant geography requires a knowledge of many other branches of botany. One must know something about the kinds of plants, for they are the units with which plant geography deals; of the way plants live, or plant physiology; of the way they are built, or plant morphology; of the way they associate with each other, or plant ecology; and of the way they evolve, or plant genetics and phylogeny. Even this botanical knowledge is not sufficient for the plant geographer. He should also know something about the origin and nature of soils, much about geology—especially its historical aspects—and much about climate and climatology.

It is obviously impossible to present all these branches of knowledge in an introduction to plant geography and equally impossible to assume that the reader is already familiar with them. We must limit ourselves to some of the simpler matters of cause and development. This is entirely possible and practicable, but the reader is warned that his own thoughts will raise questions about matters far beyond the scope of this book and often even beyond the limit of present knowledge.

Many problems in botany may be investigated experimentally and solved by inductive reasoning. For example, one might wish to determine the external conditions which affect germination of the seeds of some particular kind of plant. The seeds may be planted in various conditions of temperature,

Fig. 1.1. Spanish moss (*Tillandsia usneoides*) growing on a live oak (*Quercus virginiana*) in a park in New Orleans, La. Spanish moss is a flowering plant, related to pineapples, rather than a true moss. It roosts on other plants without taking any nourishment from them. Such plants are called epiphytes. Live oak is a general term for any evergreen oak; some species of oak are live oaks toward the southern part of their range, but are more or less deciduous-leaved toward the north. (U.S. Forest Service photo by Robert Neelands)

Fig. 1.2. Giant sequoia (*Sequoiadendron giganteum*) in the Sierra Nevada of California. This specimen, known as the Frank Boole tree, is the lone surviving Sequoia in an area logged in the 1860s. Although it is now native to only a small area at middle elevations in the Sierra Nevada, the giant sequoia is successfully cultivated at many places far from there, such as in the British Isles. (U.S. Forest Service photo by Norman L. Norris)

moisture, and light, their behavior observed, and conclusions drawn; and such experiments may be repeated as often as necessary or desired. In that way, one might discover that the seeds will germinate only in the dark, only at temperatures between 50 and 60 degrees, and even then only after they have been frozen for at least 30 days. There is practically no guesswork about such a conclusion. Every reasonable chance of error can be eliminated. Others may perform the same experiments at any time and in any place and reach the same result.

Problems in plant geography can rarely be settled in such a simple way. They may extend over thousands of miles of space or thousands of years of time. They may cover not merely a single phase of the life of the plant, such as seed germination, but every phase from the sprouting of the seed to the ultimate death of the plant. They cannot be solved or even seriously attacked by the experimental method. Like most problems in geology, they must be solved or at least approached by deductive reasoning.

In such an approach, all pertinent facts are first assembled. Second, all possible theories in explanation of the question are considered. From these various theories, one is selected which will not only answer the question under investigation but will also fit in with all the other pertinent facts. Then if *all* the pertinent facts have been considered and if the relation between the theory and the facts has been *correctly* interpreted, this one theory is the correct explanation.

As an illustration of the deduction of theories without full knowledge of the facts, it may be interesting to refer to some ideas of the German scientist, Baron von Humboldt, one of the world's foremost scientists during the first four decades of the nineteenth century. In 1802 he ascended the great mountain Chimborazo in Ecuador, more than 20,000 feet high. He was impressed both by the decrease in temperature as the altitude increased and by the corresponding change in plant life, facts which must have been observed by thousands of persons before him. He then deduced that the zones of temperature—from tropical at the foot of the mountain to arctic on the snow-covered summit—were very similar to the zones met at low elevations when traveling northward from the tropics toward the pole. From this deduction he reasoned that the vegetation of different zones on the mountain must be similar to the types of plant life through which one would pass on a trip northward into the arctic. For a century afterward this deduction remained one of the axioms of plant geography. It is now known that in spite of some similarities there is also a vast difference between the climatic zones on a tropical mountain and the

Fig. 1.3. Altitudinal zonation of vegetation in the Sawtooth Mts. of Idaho. Semi-desert shrubs in the foreground, coniferous forests on the lower slopes of the mountains, and alpine tundra toward the top. Humboldt's conclusions on the relationship between altitude and latitude are more nearly valid in temperate-zone mountains, such as these, than in the tropics. (U.S. Forest Service photo by Bluford W. Muir)

broad zones on a continent, and an equally vast difference in the plant life. One of the most important differences is the relatively slight seasonal change at upper altitudes on a tropical mountain, as contrasted to the dramatic change from winter to summer conditions at higher latitudes.

We must not ridicule Humboldt. He was a great man and a great natural philosopher, but he was not always in possession of enough facts. In his *Travels* he devotes several pages to proving to his own satisfaction that malaria is caused by the noxious exhalations from mangrove trees and other swamp-loving plants. Some of our own ideas may seem equally ludicrous to our descendants of the twenty-first century.

This does not mean that the present readers must do all the work for themselves. A great deal of the work has been done by scores of plant geographers working in all parts of the world during the past hundred years and more. We shall bring together here a number of the important facts of plant distribution as well as a number of the generally accepted explanations of these facts, and try to organize them in such a sequence that the reader may proceed smoothly and logically from one to the next. Likewise, we shall try to present these matters in such a way that any intelligent person can understand them without extensive preliminary study of botany. Nor shall we hesitate to say that we do not know in the numerous instances where good explanations are not yet available.

Reading a book is often the most practicable way of gaining knowledge, but every good book is more convincing if one can see for himself the things which the book discusses. All the simpler, more elementary facts of plant geography, such as are discussed in the first twelve chapters of this book, can be seen in one's own vicinity, as one either drives across the country or strolls through the woods and fields. Try to find for yourself actual illustrations of the subjects discussed. Conversely, observe facts for yourself and search for the explanation, either in the pages of this book or by use of your own reasoning powers.

On the other hand, the broader aspects of plant geography can seldom be seen and can never be fully appreciated without extensive travel. Your knowledge and understanding of plant geography will be greatly augmented by travel, and your appreciation of travel will be greatly increased with a little knowledge of plant geography. The grandeur of the landscape may depend primarily on rocks and mountains, lakes and rivers, but its charm, its strangeness, its picturesqueness, usually come from the plants.

Just two more preliminary matters: The two words *plant geography* and the single word *phytogeography* are precisely equivalent and both of them will be used. It is often advantageous to illustrate the subject under discussion by reference to particular kinds of plants as examples. These kinds, as far as practicable, will be common and well-known species; most of them will be natives of the northeastern United States. Naturally other examples must often be chosen from different parts of the country or even from other parts of the world. Many readers may not be directly familiar with such plants except by name.

Here is the first general conclusion for the reader to ponder; one not to be discussed in detail. It is axiomatic in nature, but it is possible that the reader has never thought about it before. *Every feature of the general distribution of plants over the world is due to the combination, in varying patterns, of the separate individual distributions of all the kinds of plants.* Think about this statement, as you read the following pages.

Our first task then is to consider the distribution of a kind of plant, not of any particular kind, but of any single kind in contrast to a group. That will be the subject of our study through the next ten chapters. When we understand some of the general principles of distribution of one kind and some of the processes and conditions which affect or determine its geographical range, we can proceed to the general distribution of groups of plants.

2/ Distribution and Range

The reader has observed that the phrase *geographic distribution* appears in the quotation from Hooker at the beginning of the first chapter. This word *distribution* has been in general use in this sense for a century or more, notwithstanding its double meaning in our language. The word may signify either the act of distributing or the result of the act. In phytogeography it always means the result. It is a long word, and to avoid monotonous repetition it is frequently replaced by *range*. While the two words are commonly used interchangeably, they are not precisely synonymous. *Distribution* tends to mean general ideas, general principles, general conditions, and this is the sense in the quotation from Hooker. Range, on the other hand, usually refers to concrete, observable facts. Notice the different shades of meaning for these two words in the following two sentences: The distribution of the American pitcher plants is remarkable. There are three genera of these plants, each with a restricted range far removed from that of the other two.

The larger animals of our vicinity—the mammals, birds, amphibians, reptiles, and fishes and many of the smaller ones—especially the insects—are gifted with the power of locomotion. They can travel as individuals from place to place to find food and shelter, to escape their enemies, or even to avoid the winter season. Even the mole, the most sedentary of our mammals, can extend his tunnels through a considerable part of our gardens. The squirrels in our towns range over a fair-sized neighborhood. The deer in our forests cover several miles. Many of the birds travel a thousand miles or so every spring and fall. For such animals we can speak of the range of an individual. We can lay out on a map, if we wish, the area over which a wren hunts for food in the summer; or on a larger map we can show its destination and route as it flies south in the fall.

None of our larger plants has any power of locomotion comparable with

Fig. 2.1. California pitcher plant (*Darlingtonia californica*), which is native to only a small area in northwestern California and adjacent Oregon. Another genus of pitcher plants, *Sarracenia*, occurs in eastern United States, especially on the southeastern coastal plain. A third, *Heliamphora*, is restricted to the mountains of Venezuela and British Guiana. (Photo by W. H. Hodge)

that of an animal,[1] but a few of them are easily moved from place to place. These are the various kinds of plants which float free on the surface of the water and are borne hither and yon by the currents and winds. Growing in the quiet waters of a pond, the little duckweeds (*Lemna* and *Spirodela*) are blown from one side to the other. If they get out of the quiet backwater of a stream and into the current, they drift down the stream for an indefinite distance. Great masses of them annually float down the Illinois River and may be seen, still cohering in large masses, far down the Mississippi on their way to the Gulf of Mexico. Water lettuce (*Pistia*), a tropical relative of the jack-in-the-pulpit (*Arisaema*), grows profusely in an inland lake in the Philippines, and myriads of detached plants float in a never-ending procession down the Pasig River through the city of Manila to meet their death in the salt water of the bay. The water hyacinth (*Eichhornia crassipes*, Fig. 2.2) also floats free on the surface and because of this habit it has become a pest in the rivers of Florida and in many parts of the tropics.

Most of our plants, on the other hand, are rooted to one spot and remain there until they die, or, if they spread by the growth of creeping stems or underground parts, their progress is but a short distance, often only a few inches per year. This habit of growth cannot properly be compared with the locomotion of animals. The ordinary plant, *as an individual*, accordingly has no geographic distribution. When we talk about the distribution (or range) of a plant we always mean the distribution (or range) of a kind of plant or of a group of plants. To express the matter with greater precision, we mean the area occupied by all the individuals of a kind of plant or a group of plants taken collectively. Thus the sugar maple (*Acer saccharum*, Fig. 2.3) is said to be widely distributed over the eastern United States; this means that individual trees of sugar maple may be found in most or nearly all parts of the country east of the Mississippi River.

Our statement that an ordinary plant, as an individual, has no geographic distribution, may be regarded as axiomatic. In geometry, an axiom is a statement universally recognized as true, on an intuitive basis, without the need for precise mathematical proof. In botany, an axiom may be regarded as a statement so simple, so obvious, and so general in its application that its truth is commonly accepted without further thought or consideration. Just for that

[1] There is one stage of a plant which may show great mobility, although it has no motility. That is, it is easily and regularly moved by other agencies but not by its own efforts. That stage, of course, is the seed (or, in non-seed-bearing plants, the spore), and the nature and effect of this mobility will be discussed in the next chapter.

reason, botanists will do well to direct a little of their thought to some of these axioms, sometimes to appreciate their relation to other botanical principles which depend on them, and in this particular instance to contrast conditions in plants, where this statement is valid, with those in animals, where it is usually not true.

The difference in this respect between plants and animals brings into animal ecology and geography some principles completely absent from plant ecology and geography. For example, many kinds of animals evince territoriality. By this we mean that a single animal, or a single pair, claims domain over a certain bit of land or water, which is its range, and fights to exclude other animals of its own kind from it. Territoriality obviously implies that the animal knows his own kind and can be aware of the presence of intruders in his own plot. The plant has no such perception, has no individual range, and cannot show territoriality. Of course, many animals do not show it and some are actually gregarious, preferring to share the same community with others of their own kind. Warblers even go further and several species of them often migrate together in a single flock.

We shall not have much occasion in subsequent chapters to refer to animal ecology and animal geography as such, although the effects of particular kinds of animals on plants and the plant community must of course be considered. The animal part of a biotic (plant and animal) community is largely governed by the plant part, rather than the reverse. Animals depend directly or indirectly on plants for food and cannot live without them. Plants, on the other hand, make their own food from raw materials drawn from the soil and the air. The total volume of vegetable material in a biotic community is ordinarily much greater than the volume of animal material, so that it is the plants which dominate the scene. Furthermore, the animals, being motile, can to some extent directly select the habitat most suitable for them. For all these reasons, it is possible (and in botanical circles customary) to consider the plant communities as such, with only secondary, if any, attention to the animals.

The animals, of course, do have some effect on the nature of the plant community. Even aside from the influence of man, grazing and browsing animals can have an important influence; many kinds of flowering plants depend on insects for pollination; birds, squirrels, and parasites eat many seeds that might otherwise find a place to germinate, etc. A full understanding of the plant community, therefore, requires a consideration of the animals. Still, the biotic factors (other than competition among the plants themselves)

Fig. 2.2. Water hyacinth filling a bald cypress swamp in Louisiana; Spanish moss is festooned on the branches of the bald cypress. (U.S. Forest Service photo by R. K. Winters)

Fig. 2.3. A solitary sugar maple tree in North Carolina. (U.S. Forest Service photo by E. S. Shipp)

Fig. 2.4. American elm tree in New Hampshire. The American elm is a characteristic tree of floodplains and similarly moist sites in eastern United States. Now seriously threatened by the Dutch elm-disease, introduced from Europe, it may eventually disappear from our forests and streets, as the American chestnut has. (U.S. Forest Service photo by W. R. Mattoon)

generally exert less obvious control over natural plant communities than the climatic or even the edaphic (soil) factors.

Returning to a consideration of ranges, we should first note that the range of a particular kind of plant may vary from very restricted to very extensive. There are a good many kinds of plants which are known from only a single locality. Most of these are from remote parts of the world into which botanists have seldom penetrated, and they will probably be found to have a wider distribution when further explorations are made. An example of such a plant is the pretty, yellow-flowered shrub *Acanthella conferta*, which for more than a century was known only from a single rocky hill in southern Venezuela. In recent years more botanists have visited this region, and the plant is now known to be fairly abundant on granitic hills along the upper reaches of the Orinoco River.

But we do not need to go beyond our own country to find excellent examples of plants of restricted range. There *was* one kind, *Psoralea stipulata*, known to exist only on a single island in the Ohio River. It was never found elswhere, has not been seen alive for many years, and is believed to be extinct. There is another with large and conspicuous rose-purple flowers, *Iliamna remota*, known only from two remarkably different places, the one an island in the Kankakee River in northeastern Illinois, the other a mountain-top in West Virginia.[2] The handsome flowering shrub *Neviusia alabamensis* was known for many years from just one station near Tuscaloosa, Alabama, although it is hardy in cultivation as far north as Boston. Recently another station was discovered in Arkansas. *Franklinia alatamaha*, a handsome white-flowered arborescent shrub, was discovered almost two centuries ago in southeastern Georgia, has not been seen growing wild since 1790, and is generally presumed to be extinct as a wild plant. The reader will readily understand, without being told, why most such plants do not have English names, although the name snow-wreath was invented by someone for *Neviusia alabamensis*. *Franklinia* has become rather well-known in cultivation and has acquired the name Franklin tree, but many other rare plants have never come to public attention.

In southern California the Torrey pine (*Pinus Torreyana*, Fig. 16.11) is known only from Santa Rosa Island and from a single grove on the mainland, where it is being carefully preserved. Readers who drive southward from Los Angeles to San Diego may be interested to know that they will pass through this reservation just before entering La Jolla. Even more limited in

[2] But see footnote 5 in Chapter 4.

Fig. 2.5. A group of quaking aspens in Utah. The conspicuous white bark with black scars is very characteristic for the species. The common name refers to the fact that the leaves tremble in even the slightest breeze. (U.S. Forest Service photo by P. S. Bieler)

Fig. 2.6. Bracken fern in a forest in Wisconsin. This ubiquitous plant is differentiated into numerous regional phases in various parts of the world, but no sharp lines can be drawn among them. (U.S. Forest Service photo by Lee Prater)

range is the shrub or small tree, Gowen cypress (*Cupressus Goveniana*), which grows only in a single patch of a few acres at Monterey, California.

From such minimal ranges there is every variation in extent up to plants which are found over a large section of our country, as the sugar maple (Fig. 2.3) and the American elm (*Ulmus americana*, Fig. 2.4) in the east or the Douglas fir (*Pseudotsuga Menziesii*, Fig. 22.64) and the ponderosa pine (*Pinus ponderosa*, Fig. 9.2) in the west, or right across North America from coast to coast, as the quaking aspen (*Populus tremuloides*, Fig. 2.5) and the paper birch (*Betula papyrifera*, Fig. 16.12), or almost throughout the north temperate zone, as the common rush (*Juncus effusus*), while a few, such as the common bracken fern (*Pteridium aquilinum*, Fig. 2.6) are almost cosmopolitan.

In the United States, good information on the distribution of trees is available, partly because they are easily observed and recognized at most seasons of the year and partly because of their commercial importance. Much observational work has been done and many important publications have been issued by the United States Forest Service and by similar agencies in many of the states. The ranges of our shrubs and herbs are less accurately known, but still sufficiently well to give a clear idea of their general distribution. Extensions of range are frequently reported by observant botanists who have found a plant growing well outside its previously known area. For example, several plants of the southern Appalachian Mountains have recently been detected in the hilly parts of eastern Ohio, and many plants of the southern coastal plain, not known before to extend northward beyond the southernmost corner of North Carolina, have been found in eastern Virginia.

There is only one way by which the range of a plant can be determined, and that is by direct observation. Some competent person must see the plant growing and make a record of the locality, and it is usually a good idea for him to collect and preserve a specimen of it so that his identification may be checked whenever desired.

If we ourselves were to start to determine the range of a certain kind of plant, let us say the New England aster (*Aster novae-angliae*, Fig. 2.7), we could travel extensively over the eastern half of the country at its blooming season in the autumn, and list every place where we saw it growing. That would be easy, since the plant is very conspicuous when in bloom, likes to grow along roadsides, and can be recognized accurately from a moving car, but it would take a long time and be very expensive. Fortunately, this work has already been done for us by thousands of botanists, both professional and amateur, who have collected far and wide over almost every square mile of

Fig. 2.7. New England aster. The species, common in moist low ground in open places in northeastern United States, is readily recognized even from a distance by the reddish-purple color of the flower heads, the other species in similar habitats being usually more bluish or white. (New York Botanical Garden photo)

our land, and their specimens are still preserved in our various colleges and botanical museums. We can determine the range of the New England aster much more efficiently by visiting these institutions. The preserved specimens normally show the place of collection, and we can make a dot on the map to indicate where each specimen was found. We would soon have a large number of dots, well scattered over the eastern half of the country, and they would give us a fairly accurate idea of the range of the plant. On the other hand, the dots would not tell us accurately whether the plant is common and abundant, and therefore an important component of the vegetation, or scarce and therefore of little importance. Conspicuous plants are often collected far out of proportion to their relative abundance.

Such work has been done repeatedly, and every book dealing with the native plants of the country states the known range of each in general terms.[3] The ranges of European plants are known even better, but for South America, Asia, and Africa the ranges of most species are known only sketchily.

Numerous books have been written by taxonomists in which they have listed and described the plants of nearly all parts of the United States and Europe and many parts of the rest of the world. Through the use of these books, we can learn the kinds of plants far more easily than we could differentiate them independently. To be sure, we may expect that additions will be made to taxonomic knowledge, even in the United States and Europe, and that some mistakes will be discovered and corrected, but for the United States these will be so slight that they will not undermine our work in plant geography, so far as general principles and general conclusions are concerned. In fact, if future experience is like that of the past, each improvement in taxonomic knowledge will make clearer the conclusions which the phytogeographer may make.

[3] Examples of such statements: For New England aster, from *The New Britton & Brown Illustrated Flora*, "Mass. and Vt. to Ala., w. to N.D., Wyo., and N.M." For tulip tree, from *Gray's Manual*, "Worcester Co., Mass., to s. Ont., Wisc., and southw." For California poppy, from *A California Flora*, "Many Plant Communities; mostly of cismontane Calif. and w. part of Mojave Desert."

Fig. 3.1. Plumed seeds of milkweed (*Asclepias*) bursting from the pod. (Photo by Lynwood M. Chace, from National Audubon Society)

3/ The Migration of Plants

How have plants attained the range which they now occupy?

Have they always been there? There are sections of the United States which are very old, to be sure, but "always" is a long time, and there is no part of the country which has not at some time in its history been under water or covered by glacial ice. Certainly, the plants which now occupy the land have not always been there.

Have they originated by evolution all over their present range? That is hardly possible. Evolution of animals and plants is a complicated matter, and we shall have to return to it later to consider some aspects of the relation of evolution to distribution. For the present we need only say that a theory of evolution throughout the present range would merely transfer our question back to the ancestors of the present species. How, then, did the ancestors attain their range? How did the first plants establish a range over land newly emerged from water or just uncovered from a mantle of glacial ice?

Or have they reached their present distribution by migration? That is by far the best solution. We can see the migration of plants going on every year in our own vicinity, and we need only multiply what we see by the thousands of years during which it has been going on to appreciate that migration is sufficient to account for many (but not all) of the facts of present plant distribution. We shall have to consider not only the way in which migration takes place, but also when it began, whence it came, how rapidly it proceeds, in what direction it is progressing, and whether it is yet completed.

The means by which plants migrate ordinarily are simple and readily observed by anyone. It will require only a brief discussion here. In some instances, it is an appropriate subject for the Boy Scouts or the nature study class. Nevertheless, it may be well to give a brief classification of the various methods used by plants and a short discussion of their relative efficiency. We can arrange the methods in six groups: wind, water, animals, man, propulsion, and gravity.

We have already noted that the individual plant ordinarily has no method of migration but remains rooted in one spot. That is true, but there is one stage in the life history of the plant which may be very mobile, and that is the seed or the fruit. It is at this reproductive stage that migration takes place.

Migration by wind is perhaps the most readily visible means. We can actually see the dandelion seeds[1] start their journey on a gusty spring day; we can see them floating through the air; we can see them swoop low enough, following the erratic movements of the air, to strike some plant or other obstacle and fall to the ground, ready to germinate, if possible, in their new home. Fruits with the general nature of a dandelion are characteristic of about one-sixth of American plants, e.g., goldenrod, aster, thistle, cattail, and many others, some less and some more efficient than the dandelion. True seeds of the same general appearance, but produced within a fruit and released at maturity, are borne by willow, cottonwood, poplar, and milkweed (Fig. 3.1), as well-known examples, and of course by many other kinds as well. Other plants may have a flat wing which serves as a sail, but is probably less efficient than the tuft of hairs of the dandelion. In pine this wing is at one end of the seed; in birch, maple, and ash (Fig. 3.2), at one end of the fruit; in catalpa (Fig. 3.3), at both ends of the seed; and in tree of heaven (Fig. 3.7), at both ends of the fruit. These examples are all fairly large and probably fairly efficient. From them, there is every gradation toward smaller and smaller wings and hairs, in relation to the size of the seed or fruit, with progressively less efficiency. The seeds of orchids and some members of the heath family, although not winged or plumed, are as fine as dust and are actually blown like dust through the air. The tumbleweeds (Fig. 3.4), of which we have several kinds, are much-branched plants nearly globular in shape. They break off at the ground when they reach maturity in the fall and then go rolling before the wind across open fields, scattering their seeds as they go. Of course they can not function among tall plants, but on the Great Plains, where they are especially abundant, they may be blown for miles. Long piles of them, dammed up behind wire fences, are a familiar sight along many western highways.

Distribution of plants by water, as it is ordinarily thought of, is not very effective, since it is available only to water plants and to those land plants

[1] The so-called seed of the dandelion is actually a fruit. A fruit, botanically, is a ripened ovary, with or without other parts of the flower attached to it; whereas seeds, one to many in number, are contained within the ovary. Thus a pea pod is a fruit and the peas within it are seeds. It is often true that a one-seeded fruit never opens to allow the seed to escape, and such is the case with the dandelion, corn, acorns, and many others. These fine distinctions, unimportant for our purpose, are vital to students of the classification and evolutionary relationships of plants.

Fig. 3.2. Winged fruits of black ash (*Fraxinus nigra*). Each fruit contains a single seed. (U.S. Forest Service photo by W. D. Brush)

Fig. 3.3. Fruits and winged seeds of northern catalpa (*Catalpa speciosa*). (U.S. Forest Service photo by W. D. Brush)

which grow so close to ponds or streams that their seeds can fall in. There are some plants of ocean shores, however, whose seeds can be carried by currents or tides and washed ashore by waves. Desirable in such seeds are light weight and resistance to penetration by water. The fruit of the coconut (Fig. 3.5), not as seen in our markets, but as produced on the trees in the tropics, is a good example. The outer layer is hard and smooth and almost waterproof. Within is a mass of fibers with air spaces, giving the fruit enough buoyancy to float the heavy nut in the center. Along tropical seashores many plants produce fruits or seeds of this general character, but naturally of many sizes and shapes. This method of migration has two notable drawbacks. First, the seeds are seldom washed far enough up on the beach to reach land where they can grow properly, and second, all such seeds are eventually killed by penetration of salt water if they are exposed to it long enough. Nor can they get from one island to another unless there are definite currents, or unless they are blown along the surface by the wind. In spite of these deficiencies, the prevalence of the same kinds of plants along the shores of hundreds of tropical islands testifies to the efficiency of the method. It is questionable how many plants (except true water plants, as the water lily or the pondweeds) we have in the United States which regularly migrate in this way. It seems probable that there are relatively few, although there is no question that many are so carried by the flood waters which often inundate the alluvial land along our rivers. Rain water, which often runs over the surface of the soil for some distance before it sinks in, must also be active in the local movement of seeds, but of course is of little or no importance in long-distance migration.

There are several different ways in which animals, with their powers of locomotion, assist in the migration of seeds. The most familiar type is the production of edible fruits with indigestible seeds. Such fruits are eaten chiefly by birds, and the seeds, still capable of germination, are voided at some distance from the parent plants. About one-eighth of all the seed-bearing plants in the northeastern states fall into this category. Scarcely second to this method is the production by plants of swamps and marshes of numerous minute seeds which merely fall to the ground at maturity. But wading birds pick up these seeds in mud which adheres to their feet and they may (or may not) carry them to the next marsh. Charles Darwin found more than five hundred seeds in the bits of mud on the feet of a wild duck. All these wading and swimming birds are known to fly during their annual migrations from one pond or marsh to another and, considered collectively, they must do an enormous business in the transport of seeds. Land birds help too, since they go down

Fig. 3.4. Tumbleweeds (*Salsola kali*, the Russian thistle) piled up along a fence in Kansas. (U.S. Soil Conservation Service photo by B. C. McLean)

to the water to bathe or drink. Next, there is a considerable number of plants collectively known as stick-tights, whose seeds or fruits are covered with hooked hairs or spines and attach themselves to our clothing as we walk through the woods in the fall. They also attach themselves to mammals and probably even to birds which come down to ground level. The American bison and the antelope, which used to roam the Great Plains and migrate considerable distances, must have been important agents in seed migration. In fact, one native plant of the region is still known as buffalo bur (*Solanum rostratum*). And lastly, there are the animals which habitually gather fruits or seeds for food and transport them some distance for storage. We think first of the squirrels which bury seeds in our lawns and forget where they hid them, or the California woodpeckers and jays which carry acorns and often drop them accidentally. Ants are probably more efficient than these creatures, since many species regularly gather seeds for food. Comparatively little is known about seed-transport by ants in this country, but in Europe there is quite a list of plants which are known to be so carried and on which the ants merely nibble at the seed coat, leaving the seed still able to germinate.

The most spectacular agent in plant migration has been man himself, because man in his migrations can and does take things with him, and among his various other items he has often carried seeds and plants, sometimes intentionally, sometimes unwittingly. Even prehistoric man took part in this, although his migrations were usually shorter and slower than those of today, and the plants which he carried were usually those actually used by him. Probably the best evidence that the Polynesians ever reached the Americas is the widespread cultivation of the sweet potato (Fig. 3.6) throughout Polynesia, for the sweet potato is strictly an American plant. In general, anthropological investigation of the origin and migration of primitive or prehistoric races is always aided by a study of the plants which they used. As a result of more recent migrations, many plants originally native to Europe now grow wild in the United States, having been brought here during the last three centuries for use for ornament. Having been assisted, as it were, in making the big leap across the Atlantic Ocean, they then proceeded to spread over the country by their own normal migration. One of the first such plants to appear was the barberry, brought to New England soon after the first colonies were established there, and later spread widely over the eastern states. One of the more recent introductions is the king devil (*Hieracium aurantiacum*), which was deliberately planted in New Hampshire and rapidly became the beautiful but pestiferous weed of eastern meadows and roadsides. Voluntary importation of plants is

76859

by no means the only way they can arrive. Their seeds come in packing material, in ballast, and as impurities in the shipments of other kinds of seeds. Within the country they travel long distances in cattle cars and even in our own automobiles.

A remarkably large number of plants of foreign origin now grow in the United States. In the northeastern states and adjacent Canada, they constitute about 18 percent of the total number of species; elsewhere, the proportions are somewhat less. All of these owe their presence here to the activities of man. Helped by human activities, some of them have spread rapidly and widely across the country. Others, after their arrival, have been left to their own resources and have spread slowly or not at all. A familiar example to easterners is the tree of heaven (*Ailanthus altissima*, Fig. 3.7), the tree which "grows in Brooklyn." By means of its winged fruits, it has spread to vacant lots, alleyways, sidewalk gratings, and even housetops. In recent years it has expanded beyond the strictly urban areas and is getting established in the groves and woodlots near our cities. Probably no foreign plant has become so completely and so successfully established, or has displaced the native plants so completely, as the wild oats (*Avena barbata*), which has occupied thousands of square miles of grazing lands and open forests in California (Fig. 22.86).

Less effective than any of the other methods described, but highly interesting to the observer, is propulsion in which the plant furnishes its own power. Fruits of various plants develop diverse sorts of internal tensions as they dry during ripening. Finally something gives way, the tensions are suddenly released, and the seeds are forcibly expelled. The most remarkable of these plants are members of the bean family. They produce pods which split lengthwise into two halves at maturity, just as do ordinary peas and beans, and each half twists or coils. Twisting is impossible as long as the two halves are united, but as drying proceeds considerable tension is set up. This tension becomes strong enough to break the connection between the two halves; each instantly coils, and the seeds scatter in all directions. In South America I once tried to preserve one of these pods unopened by picking it while it was still immature and drying it slowly by artificial heat. It was about eight inches long and two inches wide, with a very hard, thick shell. A few days later when I returned from a field trip, my Indian helper, who had stayed in camp to look after my collections, met me in a high state of excitement. "Chief," he said, "you know that big pod? I just look at him and he say 'POW!' "

The eastern American touch-me-nots (*Impatiens*, Fig. 3.8) have a small fruit which splits lengthwise into five divisions, each with an uncontrollable urge

Fig. 3.5. Left, coconut palms in Costa Rica. (New York Botanical Garden photo.) Right, a single coconut, split open. (Photo by Walter Hodge)

Fig. 3.6. Sweet potato plant. (Photo by W. H. Hodge)

to coil up. This they can not do, as long as their tips remain united. If the tip is pinched lightly, or even jostled when the pod is ripe, the five divisions coil instantly and throw the seeds as much as eight feet. There are still other propulsive mechanisms in other kinds of plants. Witch hazel and violet shoot their seeds by pinching tightly against them, just as a child shoots a slippery cherry-stone from between his fingers.

There still remain hundreds of kinds of plants which have, as far as we can see, no specialized means of migration. The fruit merely ripens and the seeds drop to the ground. And yet these plants have attained large geographical ranges.

Fully as important as the normal means of migration are the accidental migrations which any kind of plant may sometimes make. For example, a church steeple was struck by lightning in my home town one spring, and in the following summer a healthy plant of lamb's-quarters (*Chenopodium album*) grew in the exposed crevices in the bricks many feet above the ground. An American fireweed (*Erechtites hieracifolia*) made its appearance some years ago in the rubble of a bomb explosion in London. A plant of skunk cabbage (*Symplocarpus foetidus*, Fig. 4.2), normally an inhabitant of swamps, was found growing in the wet hollow of a decayed stump far from the nearest swamp. The hurricane grass of the Bahama Islands is so named because it first appeared there following a hurricane. Such spectacular instances of migration can rarely be explained; yet they are numerous in nature and, in the aggregate, must contribute greatly to the general migration of plants.

Here are four illustrations of unexpected migrations which took place on my own premises in lower New York state. A plant of Indian pipe (*Monotropa uniflora*) appeared one summer under the rhododendrons. It had never been seen in the neighborhood before. This plant has dust-like seeds. A plant of salt marshes, the marsh fleabane (*Pluchea*), appeared in my rose garden and persisted three years before dying. Here the answer is easy, for the garden had been mulched with salt hay. A plant of monkey flower (*Mimulus*) appeared in the same rose garden, bloomed and bore a good crop of seeds, and from that year numerous seedlings came up in many places in the garden. A plant of great blue lobelia, normally an inhabitant of wet soil, appeared in a garden bed and bore a good crop of seeds. From these, a patch of lobelias about fifteen feet long gradually developed, with scores of individual plants which thrived greatly. Although they produced thousands of seeds, not one seedling was ever found elsewhere in the garden. Here are four plants which migrated into my garden, three of them by unexplained means and from unknown sources; one of these, the monkey flower, was able to spread still farther by

Fig. 3.7. Tree of heaven (*Ailanthus altissima*). Left, leaf and female flowers; right, winged fruits. (U.S. Forest Service photos by W. D. Brush)

methods also unknown. Could they have come in on my own clothing, after a hike through the woods? Perhaps, but none of them has seeds especially adapted to this method of distribution.

There are, in fact, many instances of migration which could not have been made by normal means within the time available for them. In the eastern states, for example, there is scarcely a brook or roadside without its several colonies of touch-me-not (Fig. 3.8), often also called jewel weed. Even a minute wet hollow in the woods, far removed from a brook or pond, will be colonized by them if the shade is not too dense. No one knows how they get there, because the explosive discharge of the seeds covers only a few feet and there may be a hundred times that distance between such an isolated colony and its nearest neighbors. If the seeds are ordinarily discharged up to eight feet, about 650 years would be necessary for the plant to travel a mile, and a journey of 300 miles would require about 200,000 years. Yet these plants occur naturally far north of the glacial boundary in land which has not been free from glacial ice more than a few thousand years.[2] To reach these northern stations in a much shorter time, the plants must have made frequent and fairly long migratory leaps by unknown or accidental means. Our best guess is that they traveled in mud clinging to birds' feet.

How far do seeds migrate at a single leap? For most kinds of plants; no one knows, nor can useful conclusions necessarily be drawn from direct observation. The winged seeds of silver maple (*Acer saccharinum*) have been seen to travel as much as 800 feet on a windy day, but it is easily observed that the great majority of them reach the ground only a few feet from the parent tree. Yet, the few which are carried the maximum distance may be more effective for long-distance migration of the tree than the many which make the minimum journey. I have seen a colony of reed (*Phragmites communis*, Fig. 3.9) growing on top of a rock more than a mile from the nearest colony and have had fairly satisfactory evidence that the cattail (*Typha*, Fig. 13.7) may travel as much as two miles in one jump.

One might suppose that plants with edible fruits and indigestible seeds would be carried long distances by birds. However, this opinion is probably seldom correct. At the time the fruits are ripe, birds are fairly well settled in one area. Also, the digestive processes of birds are rapid and the seeds are voided so soon that long-distance migration is certainly unusual. Yet, how can

[2] It is only about 11,000 years since the most recent ice sheet reached its fullest extent in North America and began to recede. Northward from the glacial boundary the ice-free time is of course progressively shorter.

Fig. 3.8. Touch-me-not (*Impatiens*). The curious flowers are very showy and distinctive. The most common species in the eastern United States has the flowers orange-yellow with brown spots. A less common species has the flowers pale yellow with brown spots. (Photo by N. E. Beck, Jr., from National Audubon Society)

Fig. 3.9. Reed (*Phragmites communis*). This species grows in moist places over a large part of the world. It is tolerant of salt and is especially common in brackish lowlands near the coast, as for example in the marshes toward the northern end of the New Jersey Turnpike. (Photo by H. W. Kitchen, from National Audubon Society)

we account for the distribution of crowberry (*Empetrum*), which grows in the northern parts of North America and reappears in Patagonia? A constipated tern, perhaps?

Water currents carry fruits and seeds long distances. Many marsh or water plants have seeds covered with corky tissues which keep them afloat, but all of them become waterlogged in time, and, if they do not find suitable lodging, they sink and perish. Seeds of South American plants, washed into the ocean there, have been carried thousands of miles by the Gulf Stream and stranded on the shores of Norway, but of course they were all dead when they got there.

Of all the natural means of migration, the most effective is doubtless the transport of seeds by water birds in the mud on their feet. It probably accounts for the almost cosmopolitan distribution of such plants as the common rush (*Juncus effusus*) as well as the wide range of many other marsh plants. In the last few centuries, man has become the most efficient of all agents of plant migration through his commerce, trade, and travel. If a tire on your car picks up the prickly fruit of a puncture weed (*Tribulus terrestris*) in California and it does not fall off until you reach Maine, you have certainly given the seed a free ride of some three thousand miles. Actually this can happen.

It is harder to account for the migration of plants with large, heavy seeds—such as oak, walnut, or hickory. Apparently the only way they can travel is by rolling a short distance. Squirrels and other rodents gather them for food and may carry them somewhat farther away. Lacking the means of long jumps by accidental methods, these trees are compelled to travel by a connected series of short steps.[3] Nevertheless, these trees have been able to migrate far north of the glacial boundary in the past few thousand years since the ice disappeared, and presumably they must have made the journey by steps no longer than the foraging trips of squirrels. Oak trees have reached northern Michigan since the last glacial advance. If the distance from the glacial boundary is 500 miles, the available time 11,000 years, and the minimum age of an oak, when it first bears acorns, is 30 years, then the minimum average length of each step in the migration comes out to one and one-third miles—much more than the normal range of a squirrel. There are reports that passenger pigeons ate acorns and later disgorged them. Ornithologists do not believe this, but if it were true it would give the oaks much greater migratory ability—in the past, that is.

[3] In the Middle West isolated groves of forest trees are, or were, well separated from the main body of the forest by intervening prairie. Such groves were composed of trees distributed by wind or by birds, as ash, maple, elm, hackberry, red mulberry, and wild cherry. They did not contain oaks or other nut trees.

Fig. 3.10. Osage orange. Left, male flowers and young leaves; right, fruit and mature leaves. (U.S. Forest Service photos by W. D. Brush)

The heavy, globose fruits of the Osage orange (*Maclura pomifera*, Fig. 3.10) fall to the ground and soon decay. Apparently no bird or native mammal ever eats them or carries them away. Yet the Osage orange has made at least one spectacular migration. During one of the glacial periods which affected the eastern United States, part of Pennsylvania, almost all of Ohio, Indiana, and Illinois, and all the lands north of these states were deeply covered by ice. Osage orange must have been living at the time somewhere south or more likely southwest of this ice field. The glacial advance was followed by an interglacial period probably warmer than the present time, and during that time the Osage orange migrated north or northeast as far as Toronto. How much farther north or east it may have gone is not known, nor do we know whence it started or the route it followed. The tree made a journey of at least 300 miles and probably considerably more.

All these long migrations by plants so poorly equipped for travel are best explained by the use of accidental means of transportation, movement by strange and unusual means which we seldom see in action but which from time to time probably assist nearly all kinds of plants in long-distance migration. Also we must consider the matter of available time. Plants have had, not decades or centuries, but thousands or even tens of thousands of years to carry on their migrations, and even short steps become wonderfully effective if continued long enough.

Whither do seeds migrate? That is also a difficult question to answer by direct observation, but we can get some idea of the matter by noticing where young plants develop. It then becomes obvious that most seeds come to rest on the ground comparatively close to the parent plant, and that the number which fall at a distance decreases more rapidly than the distance increases. Some persons have attempted to reduce this statement to a mathematical formula, and, as a matter of fact, there are instances in which the number of seeds, as measured by the number of seedlings next year, seems to vary inversely as the square of the distance. That is, if there are fifty seedlings per square yard twenty feet away from a maple, there should be only two a hundred feet away. The distance is five times as great, the square of 5 is 25, and the number of seedlings should be 50 divided by 25, which is 2. It is very doubtful, however, if such a formula will hold for many plants, and equally doubtful if any statement of general application can be made. And if any general rule could be determined, it would obviously apply only to migration by the normal method and could not take account of the numerous accidental migrations which may take some of the seeds to much greater distances. In the dust storms associated

Fig. 3.11. American elm. Winged fruits and young leaves. Each fruit contains a single seed. (U.S. Forest Service photo by W. D. Brush)

Fig. 3.12. Fireweed (*Epilobium angustifolium*) in Jackson Hole, Wyoming. The plant gets its common name from the fact that it is often one of the first invaders after a fire. Its plumed seeds float easily in the wind, giving it an advantage toward the early exploitation of any newly available site. (Photo by John H. Arnett, Jr., from National Audubon Society)

with the great drought of 1934, vast numbers of seeds of many kinds of western plants were carried to the eastern United States and even into the Atlantic Ocean.

It is probably unnecessary to state that the parent plant has no control over the migration of its seeds. Of the various external agents which provide the motive power after the seed is detached from the parent, only two show any indication of a definite destination. Water currents regularly flow in one direction. Seeds which float downstream in a river can wash ashore only along the banks of the same river. Water birds, which pick up seeds in the mud adhering to their feet, usually fly only from one pond or swamp to another. They are not apt to visit an open field or a dense forest. Birds which eat fruits and void the seeds at a distance usually show a distinct preference for one type of country. Forest birds are not apt to visit grasslands nor do birds of open country forage in the forest. The transportation of seeds by birds therefore tends to be into a place more or less similar to that where the parent plant grew.

Of course, many exceptions can be found to everything said in the preceding paragraph, and about the best general statement that we can make is that seeds migrate anywhere and everywhere. Since it is difficult to observe the destination of migration directly, it may be equally difficult at first to appreciate how widely, how generally, we may almost say how universally, seeds are scattered. Let us try to present the matter by a few examples. If you have a well kept lawn, a concrete walk leading from your front door to the street, and a mature elm (Figs. 2.4, 3.11) nearby, you may have noticed that hundreds of young elms appear at the very edge of your walk, but none or only a very few in the lawn. If you look carefully in the lawn, you will find great numbers of elm seeds there too, but they either fail to germinate or the seedlings are cut off by the first trip of the lawn mower. If elm seeds, which are big enough to be seen easily, migrate all over the lawn, you can expect that other seeds, too small to be found among the grass, have migrated there also. I once watched a spot in a large lawn from which the sod was removed in the spring for transplanting elsewhere and on which a building was to be erected later. In the interval, twenty-nine kinds of plants sprang up in the bare spot, not one of which could be found in the undisturbed sod which surrounded it. That one spot was otherwise no different from the rest of the lawn and in every other spot of similar size there must have been just about as many different kinds of seeds which had immigrated unseen and unsuspected but had failed to grow.

In the eighteen-seventies a forest plantation was established on the grounds of the University of Illinois. It was remote from the nearest natural forest and on black prairie soil which had previously been in cultivation. During the next twenty-five or thirty years about a hundred kinds of native woodland plants established themselves in it, many of which were not known to occur naturally within ten miles.

But this lawn and this forest plantation, you may say, are an artificial sort of place, created and controlled by man. Is seed distribution equally effective in all places? For an answer, we go to the south shore of Lake Superior, where we may still find beautiful forests of beech and maple in which the lumberman's axe has never swung. We observe the plants growing in the dense shade of the forest floor: trillium, baneberry, maidenhair, shield fern, foam flower, and many others. Among the common plants of the region which we do *not* see are fireweed and wild raspberry. Then we find a place where a veteran tree of the forest, perhaps four hundred years old, has fallen and made an ugly gash in the canopy through which sunshine can penetrate. Already these two sun-loving plants have immigrated, and we find young plants of the raspberry and a few mature fireweeds (Fig. 3.12) in bloom. They had no way of perceiving and choosing this spot. Surely they have migrated everywhere in the forest—the raspberry by its indigestible seeds; the fireweed, with its plumed seeds, by the wind. We can not find these seeds, but in this particular sunny spot the seeds can grow, and we demonstrate the migration by its results.

The conclusion is inescapable that all parts of the land are regularly planted, by migration, with great numbers of seeds of many kinds of plants, arriving from various sources and by various methods. This is a fact which should be remembered. We can add an important corollary to it: The number of seeds which arrive tends to vary in inverse relationship to the distance they must travel.

We must be reasonable, though, and not expect too much of plants or be too impatient for results. The wood anemone grows in wet woods about a quarter of a mile from my home. In my own woods there are none. Why doesn't the plant migrate into mine? There are several reasons. There are less than a hundred plants in the existing colony; each plant produces a single flower, and some, not all, of the flowers mature a few ripe achenes. (An achene is actually a fruit containing a single seed, but it never opens and it behaves just like a seed.) These achenes have no method of long-distance migration, and a movement of a quarter of a mile will have to depend on some lucky accident. Now what is the chance of one achene of this little plant getting into my woods? Would you say it is one in a thousand, or one in a million?

Name your own odds. Next, suppose that this little colony continues to live and my woods continue to exist for a thousand or even a million years. Then, if we can wait that long, there is a good chance that the wood anemone will appear. Similarly, if you have a pond on your place and want a water lily in it, you do not need to buy one. Just encourage birds to frequent the pond, relax and wait patiently. Sooner or later you, or your heirs, will get your water lily!

Now what are we to learn from our observation of the facts of migration? First, that all plants can and do migrate, although with various speeds and to various distances. Second, that all parts of the world are regularly planted, often unsuccessfully, with great numbers of seeds. Third, that the effect of short migrations depends on the time during which they are continued. Fourth, that the chance of a lucky accidental migration increases in proportion to the time available. And last, that we must never forget the importance of time in phytogeography.

Fig. 4.1. Black spruce swamp in Minnesota, with birches on slightly higher and more open land in the foreground. (U.S. Forest Service photo by F. H. Eyre)

4/ The Principle of Discontinuous Distribution

In the preceding chapter a deliberate attempt was made to present, as vividly and effectively as possible, the extraordinary potentialities of plant migration. Nothing was exaggerated. In fact, much greater claims for the migratory ability of plants might have been made without stretching the truth. We trust that the reader is convinced that plants can and do migrate almost anywhere and everywhere. If he is not convinced by what he has read, he certainly can be by his own observations.

Now, having set up this idea, we shall proceed to discuss some of the conditions which greatly limit the effectiveness of migration. It all condenses to a single short statement. Travel alone is not sufficient; the seed must be able to germinate and grow successfully. If it can not or does not, its migration has been futile.[1] A consideration of this will bring up several topics which we need to discuss, and we shall take up in this section one of the simplest, the matter of discontinuous distribution.

Two types of discontinuity can easily be noted for every kind of plant and in every locality, while a third may be detected for some kinds.

First, one kind of plant, under natural conditions, very rarely occupies a piece of ground to the complete exclusion of all others. One may, but probably will not, find a corn field or a garden bed occupied only by corn or by lettuce, but that is not a natural condition. Even then there are usually plenty of weeds to fill up a part of the space and share the ground with the cultivated plants. One may find in our northern forests fairly extensive areas where the only kind of tree is black spruce (*Picea mariana*, Fig. 4.1), but there are various kinds of shrubs and herbs growing beneath it. In the eastern states, about the nearest

[1] There are analogous facts in human history. Before the colonization of Jamestown in 1607 and Plymouth in 1620, several attempts had been made to establish permanent settlements in America along the shores of the North Atlantic. In each of them, the migration across three thousand miles of water was successful but the colony failed.

approach to a pure stand of a single kind of plant is a cattail marsh (Fig. 13.7), in which other species are often reduced almost to the vanishing point. A bog forest of arbor vitae (*Thuja occidentalis*, often called also the white cedar) is sometimes so dense that no plants grow beneath it, at least over a limited area. The same is true of a mangrove swamp in the tropics. Such conditions as these are exceptional, and the normal state in natural vegetation is the joint occupation of the land by a number of kinds of plants mingled in different proportions. In most types of plant life in the eastern states, fifty to a hundred different species of plants will be found sharing the space in any single acre. Some of these will be represented by only a few individual plants or even by a single one. Others will be much more abundant,[2] but no one kind will have a monopoly of the space.

Second, even after allowance has been made for this first type of discontinuity, it may easily be seen by field observation that no kind of plant has an unbroken range over any considerable expanse of land.

If one drives along a country road in the northeastern states in early spring, he will see almost every bit of wet woods and some wet meadows colonized by the skunk cabbage (*Symplocarpus foetidus*, Fig. 4.2), which at this season becomes very conspicuous with its large leaves. It does not occupy all of the wet woods; it must share the space with red maple, alder, marsh marigold, and many other kinds of plants. The boundaries of each colony are quite distinct, and from the boundary over the higher and drier ground to the next colony may be any distance from a few yards to possibly a mile or more. In other words, the local range of the skunk cabbage is interrupted or discontinuous.

If we examine the sandy shore of an inland lake in the region of the Great Lakes, we shall find on the wetter beaches—especially on the west side of the lake or in sheltered coves where wave action is slight—the small, creeping, yellow-flowered silverweed (*Potentilla anserina*). It is a common plant in such places and may be found on the shores of hundreds of lakes and ponds and even along the flat beaches of the Great Lakes themselves. It does not occupy the entire shore (our first type of discontinuity), but neither is it found anywhere else. If we wish to see other colonies of it, we must go to the next lake.

[2] This matter of rarity and abundance is one of the most baffling and neglected problems of plant ecology. It is of importance to the plant geographer because the general aspect of the landscape is determined primarily by the kinds of plants, and those kinds represented by large numbers of individuals are naturally more important in producing the general aspect, other things being equal, than the rare species.

Fig. 4.2. Skunk cabbage, showing inflorescence and young leaves. (Photo by W. H. Hodge)

Fig. 4.3. Leatherleaf bog in Michigan; red pine (*Pinus resinosa*) on slightly elevated places in the bog. (U.S. Forest Service photo by F. H. Eyre)

If we examine a peat bog in the same region, we shall probably find much of the flat surface of the bog occupied by a dense growth of a small evergreen shrub, the leatherleaf (*Chamaedaphne calyculata*, Fig. 4.3). It is one of the most characteristic plants of our northern bogs but, like the silverweed on the sandy shores, it is not found anywhere else. To find a second colony of it, we must go to another bog.

Not all kinds of plants occupy as small a proportion of the total area of the land as do these three species. In the Great Lakes region, before the destruction of the forests by lumbering, we could leave the shores or bogs, enter the drier uplands, and there travel for miles through a continuous unbroken forest of white pine (*Pinus Strobus*, Fig. 4.4). But regardless of the direction we took, we would sooner or later come to the margin of the white pine forest and pass out of it into a forest of a different type in which the white pine did not grow. Similarly, before the prairies of Illinois were put under cultivation, one could have traveled for more than a hundred miles along the Grand Prairie and, by steering a tortuous course to avoid the wetter places, he could have been surrounded all the way by a luxuriant growth of the tall bluestem grass (*Andropogon Gerardi*, often also called *Andropogon furcatus;* Fig. 22.51). Yet he would eventually have reached the margin of it, because the Grand Prairie was completely surrounded by forests in which the bluestem did not grow.

The difference between the white pine and the bluestem, on the one hand, and the silverweed, leatherleaf, and skunk cabbage on the other, is merely one of size, not of character. All five of them, and all other plants as well, are located in detached areas, varying in size from a few square yards to many square miles and as irregular in shape as the pieces of any jigsaw puzzle or as any congressional district.

The reason for this second type of discontinuity is so apparent that it demands merely the briefest restatement, although it will require considerable elaboration when we begin to understand some of its implications. The environment varies from place to place, and some features of the environment, especially the nature of the soil and the amount of water in the soil, often vary greatly within a short distance. Every kind of plant requires certain conditions for its growth; and, if these conditions are not available, the plant simply will not grow. The boundaries of the detached areas described above always indicate a change in the environment, and this change is just as effective in keeping each species of plant within its own restricted area as the woven wire fence which keeps chickens in a chicken-yard. This simile has been chosen purposely, because sometimes a chicken does fly over the fence!

Another term which we shall need to use frequently should now be introduced. A *habitat* is an area of ground, small or large in its extent, over which the environment is essentially uniform.[3] Each habitat is therefore bounded by a change in the environment. Since the environment is so nearly uniform within the habitat, any kind of plant which can grow there may, and often will, grow throughout its extent, but will not pass beyond it into a different habitat. Our second type of discontinuity may accordingly be explained by stating that a kind of plant may grow, within its total range, only in the habitats which are favorable to it.

But does a plant *always* grow in every habitat suitable to it and over the *whole extent* of the habitat? The answer is emphatically no. As previously stated, plants attain their range by migration. Possibly the plant is now on its migratory way and has not yet arrived. Possibly this particular habitat is surrounded by other and different habitats so large that the plant is unable to migrate across them and will not arrive until some lucky accident of migration brings it. Possibly it has only recently arrived and has not yet had time to spread over the whole extent of the habitat. Possibly it is meeting with such strenuous competition from other plants that only a few individuals have a chance to grow.

These questions, and hundreds of others like them which constantly present themselves to the thoughtful phytogeographer, can be answered only by patient observation in the field and by patient and logical thought. Then for some questions we may find answers which appear to be correct; for others we may reach plausible theories, and for still others we may never find a clue. But always remember that the range of a plant is *attained* by the process of migration and that the attainment of anything takes *time*.

The third type of discontinuity is less common and far more difficult to explain. It relates to the existence of a species of plant in two or more widely separated areas. One rather simple example has already been mentioned: *Iliamna remota* lives in two such areas, one a small island in northeastern Illinois, the other a mountaintop in West Virginia. The difference in environment is great; the distance between them, as the crow flies, is about four hundred miles. There are several plants which live in the deserts of the southwestern United States and adjacent Mexico, are completely absent from the tropics to the south, and reappear in the deserts of Chile. Our tulip tree (*Liriodendron*

[3] The concept of uniformity is here treated according to the requirements of the case. A habitat is uniform in providing the necessary conditions for the plant under consideration. A difference unimportant to one kind of plant may be important to another.

Fig. 4.4. White pine in New Hampshire. This "Lone Pine," one of the famous landmarks of the Androscoggin Valley, has been cut. It disappeared in July, 1956, victim of persons unknown. (U.S. Forest Service photo by Lee Prater). Inset: Leaves and cone of white pine. (U.S. Forest Service photo by W. D. Brush)

Tulipifera, Fig. 4.5), which has a wide distribution in the eastern states, also lives in China.[4] The strawberry tree (*Arbutus unedo*) is primarily a native of southern Europe but occurs also in western Ireland. Widely separated areas of distribution are often described by the adjective *bipolar*. The tulip tree and the crowberry (mentioned in Chapter 3) are examples of bipolar distribution. But white pine would not be, although it has numerous scattered colonies well to the south of its general boundary, nor would *Iliamna remota*, nor would the big tree (*Sequoiadendron giganteum*) which lives in a number of isolated groves in the Sierra Nevada. However, the difference between the distribution of these latter three plants and the tulip tree or crowberry is merely one of degree, and we can assume that the distributions of the tulip tree and crowberry are merely the result of more remote causes.

There are satisfactory, or at least plausible, explanations for some examples of this third type of discontinuity, but for others, as *Iliamna remota*, we have as yet not even the vaguest idea.[5] Many of them are important in furnishing clues to the distribution of species in prehistoric times, and we shall have occasion to refer to some of them later.

Coming now to the subject of maps, we can understand that no map of the range of a species is completely accurate, because of the first two types of discontinuity. The degree of possible accuracy is determined by the scale on which the map is drawn. On a small scale map, range is usually shown as a continuous unbroken area or is bounded by a single closed line. On a larger scale, outlying colonies may be shown, or areas from which the plant is missing may be indicated if they are large enough. On a detailed map of a small area each separate habitat may be shown, while a chart of a single habitat or part of a habitat may indicate each individual of a species. All maps showing the distribution of a species must therefore be regarded as only approximations to the truth. The larger the map, the closer the approximation can be.

In summary: A kind of plant very rarely has a monopoly of an area, but must share it with others. Even within this limitation, a plant does not have a continuous range over a large area, but is restricted to certain habitats. And third, some plants have two or more areas of distribution so far apart that they can not easily be explained by mere differences in habitat.

[4] The Chinese tree is very slightly different from the American one and is sometimes treated as a closely related but distinct species. There are no other species of *Liriodendron*. See also footnote 5.

[5] Nevertheless, it may be mentioned that some botanists regard the West Virginian plant as belonging to a different species which they call *Iliamna Corei*, so named in honor of Dr. Earl Core who discovered it. While this does away with one problem, it raises another: How did these *two* species get this sort of distribution when all their relatives live in the far west?

Fig. 4.5. Tulip tree. This species betrays its tropical ancestry by continuing to produce new leaves throughout the growing season, in the manner of tropical trees, shedding the older leaves apace. (U.S. Forest Service photo by Lee Prater)

Fig. 5.1. Mullein plants. Throughout most of the United States, mullein is associated with empty beer cans as a marker of "civilization." (Photo by A. W. Ambler, from National Audubon Society)

5/ Some Considerations of Seed Production

There are many species of plants which bear only a single crop of seeds. In our country most such plants are annuals; they live only a single season and die promptly after the seeds are ripe, or continue to bear seeds until killed by frost. Most weeds of our cultivated lands and gardens are annuals;[1] in fact, plants could hardly be very successful weeds unless they bloomed the first year and spread easily by their seeds. Native annuals exist in the eastern states, as the touch-me-not (Fig. 3.8), but they constitute only a small part of the total plants growing there. The proportion of annuals increases as we move westward into areas of drier climate and reaches its maximum in the southwestern deserts. A small fraction of our plants, most of them weeds of vacant land and roadsides and mostly natives of some foreign land, are biennials, blooming and ripening their seeds during the second season and then dying.[2] The common mullein (*Verbascum Thapsus*, Fig. 5.1), a native of Europe, is a familiar plant of this type.

The great majority of plants in the moister northern, central, and eastern states and in moist or wet climates throughout the world are perennial, that is, in all regions where forests predominate. Most of them, after reaching

[1] Of course there are perennial weeds too, and some of them are pernicious, as the hawkweed (*Hieracium*) and oxeye daisy (*Chrysanthemum Leucanthemum*) which infest our meadows. Even trees may be weeds, not only in the tropics but also in our latitudes, where young plants of Norway maple, white ash, tree of heaven, and black locust spring up abundantly in our gardens.

[2] In other countries there are various kinds of long-lived plants which bear only one crop of seeds. In southern Asia many species of *Strobilanthes* produce woody stems which live several years, blossom simultaneously, ripen their seeds, and die. The beautiful talipot palm of Asia reaches a large size and a considerable age without flowering, but eventually produces several huge clusters of flowers, ripens its fruits, and dies. Most familiar to us is the American century plant (*Agave americana*, Fig. 5.2) which for many years (but scarcely a whole century) stores up reserve food in a basal cluster of large leaves, then sends up a tall flowering stem, ripens its seeds, and dies.

maturity, which may require one to many years, bear crops of seeds every year or nearly every year as long as they live.

There are also many plants which produce but a single flower each year, as the trillium, bloodroot (*Sanguinaria*, Fig. 13.5), wild ginger (*Asarum*, Fig. 14.14), and May-apple (*Podophyllum*, Fig. 5.5). Here again the number of seeds produced by each plant is limited. All of these plants are perennial, so that during their whole lifetime they may produce quite a respectable number of seeds. It is nevertheless noteworthy that a large proportion of the flowers on these plants, as well as on a great many other kinds with numerous flowers, never ripen any seeds at all. Everyone knows that most of the huge number of flowers on an apple tree never ripen their fruits, and that in general the yield of our crops fluctuates up and down from one season to another. It is the same with our wild plants. Many flowers do not produce fruit even in the best seasons, and in poor seasons a still smaller fraction of the flowers are able to mature.

This condition is not true of annuals, the very existence of which depends on the regular production of seeds,[3] and it is often surprising how they practically always bear their seeds even under the most adverse conditions. A seed of the western pennyroyal (*Hedeoma hispida*, not the common eastern pennyroyal, which is *Hedeoma pulegioides*) germinated one spring in scarcely more than a pinch of soil on top of a flat rock. By early June, when I discovered it, it had given up the struggle and died of lack of water at a total height of 7 millimeters (your lead pencil is about 8 mm. in diameter). Nevertheless, it had dutifully produced, as its last act, two flowers and ripened eight seeds.

From these plants with comparatively small seed production, we can find every variation to plants which produce huge numbers, provided the conditions are favorable. Those produced annually on a walnut or hickory may be numbered by hundreds; those of a Norway maple run into the thousands, on an ash or elm into the tens of thousands, and on a big cottonwood probably into the millions. Weeds are notorious as successful seed-bearers. A single large plant of tumbleweed (*Amaranthus albus*) was brought into the laboratory at the University of Illinois and allowed to dry out until the seeds were ripened and all shattered loose. The number was then estimated by counting and

[3] One might suppose that the continued existence of annuals would be totally dependent on the production of a crop of seeds each year, but this overstates the case. Most of the seeds produced in a given year will ordinarily germinate the following year, but in many species some of them remain dormant no matter how favorable the conditions for growth. These germinate in progressively smaller numbers over a period of several years. The gardener's aphorism, "One year's seeding is seven years' weeding," has a solid foundation in fact.

Fig. 5.2. Century plant in Arizona. (U.S. Forest Service photo by Bluford W. Muir)

weighing a thousand and also weighing the entire crop. The total number exceeded three million. Even this huge number is possibly exceeded by some orchids, in which the seeds are microscopic in size.

So much for the idea of the great seed-productivity of plants. We come next to a new thought, which may prove more difficult to understand.

In almost every natural area of vegetation, the total bulk of plant life is as great as the habitat can support. This may seem at first incredible, but it is nevertheless true, subject to a few easily understandable exceptions. Bare areas may be formed by natural forces, such as an avalanche destroying all vegetation on a mountain side, or a tornado knocking down all the trees in a forest, or a flood forming a deposit of bare silt along a riverbank. Such areas remain bare only a short time. Immigration of plants begins immediately, and in due time the maximum plant cover is again established.

We can most easily see and appreciate this principle of maximum coverage in our cultivated crops. In Iowa, with a soil and climate beautifully adapted to corn, the farmer may harvest a hundred bushels from each acre, while in other sections of the country the yield may be only a third as much or the successful growth of corn may be impossible. Corn is a crop of the grain only, not including the stalks and leaves, but the same condition is also true for hay, in which almost the entire plant-cover is harvested. It too varies from good to poor, and the total crop, no matter how good or how poor, represents the best that can be done under the circumstances.

We often read about the so-called marginal lands, where the environment does not permit the consistently successful cultivation of any ordinary agricultural crop. For some of them, experts advise the abandonment of agriculture and the reforestation of the land, as along the northern border of the United States; or the restoration of range grasses and the raising of cattle, as in some of our western states; or a change from field crops to dairying, as in parts of the eastern states. In these circumstances, the particular features of the environment which set a limit to productivity are the temperature and the length of the growing season, the amount of available water, or the character of the soil. These as well as other features of the environment also set a limit to the productivity of the natural vegetation. In our southwestern deserts, for example, where short-lived annuals are so numerous, the controlling factor is primarily the water-supply. In a dry year, comparatively few seeds are able to grow successfully and the crop of annuals is limited. In a year with more than average rainfall, more annuals are able to grow and the desert is gaudy with great displays of flowers.

Fig. 5.3. Effect of moisture differences on plant cover. Note the relatively dense and luxuriant cover immediately along the highway, where the plants get runoff-water from the road, in contrast to the sparser vegetation, with large bare spaces, away from the road. (U.S. Forest Service photo by Rex King)

Fig. 5.4. Young forest of sugar maple, with large numbers of one- or two-year-old seedlings carpeting the ground. (U.S. Forest Service photo by Lee Prater)

In the eastern forests there are two important limiting factors, water and light, which determine the crop of herbaceous plants on the forest floor. The trees usually form a continuous canopy overhead and receive the maximum light. On the ground beneath, the amount of light is greatly reduced and herbs and shrubs are able to grow only to the extent that light is available. Every easterner has seen for himself the dense growth of shrubs and herbs which springs up along a roadside in the forest or in a spot where a little cutting has been done. Also trees are often surface feeders and their roots so exhaust the water from the upper layers of soil that little is left over for herbs and shrubs. Professor Toumey of Yale University isolated small areas of forest soil between but excluding the trees by the simple method of digging a narrow trench around the area and then filling it again. This cut all the tree roots which had been entering the area from the neighboring trees and freed the herbaceous plants from their competition for water. Immediately there was an extraordinary increase in the amount of herbaceous vegetation.

This general statement that vegetation is regularly at its maximum in every natural area is true, but it deserves, and may require, a little meditation by the reader. As soon as he is convinced of its truth, he will be able to appreciate our next general principle, that the successful germination of a *single seed* and the growth to maturity of *one single seedling* by each individual plant will be sufficient to maintain that species of plant in its original numbers.[4] If two seedlings should grow to maturity, the number of individuals would double in each generation, and in comparatively few years there would be no room for any other plants at all. If the parent plant is a long-lived perennial and does not bear seeds until it reaches a considerable age, as an oak, the effect would be the same, but more years would be required to reach it. Again, if all the numerous seeds produced by a plant during its lifetime were to grow successfully, the same result would be achieved in a short time. The little wood anemone, mentioned before, bearing about 10 seeds per year, would have a progeny of a thousand plants in three years, a million in six, and a billion in nine. Almost certainly there are not a billion plants of it in existence in the whole country and never were. The big tumbleweed which produced three million seeds occupied about a square yard of ground. If all the plants grew to equal size, its progeny next year would occupy three million square yards or about a square mile, and in the third year tumbleweeds would cover

[4] Most of our plants, unlike mammals, have both sexes on the same individual. A little reflection will show that no very serious modification of the principle is required to cover plants in which the sexes are on different individuals.

the entire United States. Such a conception is, of course, an absurdity, but it serves to emphasize the truth that a great excess of seeds is regularly produced by plants.

Since only a single seed from each plant is destined, on the average, to grow successfully to maturity, it is obvious that the fate of seeds is exceedingly precarious. Many are eaten by insects or other animals. Many decay before germinating. Many fall on spots where germination is impossible. Still, for most kinds of plants, far more seeds than the single one required actually do germinate and appear above the ground as seedlings. The struggle for existence then continues. Many of the little plants die for want of water or light, unable to compete with the larger and stronger plants already established or with those of their brothers who got two or three days start. Many fall a prey to disease or are eaten by animals, and finally the whole crop of young plants is reduced to a single mature individual.[5]

Observations in a forest of sugar maple (Figs. 2.3, 5.4) have yielded some striking figures to illustrate this principle. The sugar maple bears seed abundantly and apparently every year, and its seedlings appear in the forest by the million. A fair estimate, based on actual count, is about ten seedlings to every square foot of the forest floor. If a mature tree is assumed to live 350 years, and this figure is approximately correct, and to occupy about 400 square feet, then about 4,000 young trees spring up beneath it every year, and during its total life no less than 1,400,000 of them. We need not try to guess how many other seeds failed to sprout. Of these seedlings, the vast majority succumb during or at the end of their first year, and only 70,000 persist during the second year; 1,400 live to be ten years old; 35 grow to be tall slender saplings, badly suppressed by the dense shade, and dying at about fifty years of age when still barely more than an inch in diameter; two attain an age of 150 to 200 years and a height of 60 to 80 feet, and one of these becomes a forest veteran.

A somewhat different manifestation of the same general principle is illustrated in hundreds of woodlots in the eastern and central states by the common May-apple (*Podophyllum peltatum*, Fig. 5.5). The plant spreads by branching underground stems and forms crowded colonies, circular in general shape.

[5] This statement always pertains to average conditions. It is perfectly obvious that the yellow hawkweed, which is becoming such a pest in eastern meadows and roadsides, has been increasing its numbers at a nice rate of compound interest. But every new colony of it occupies space which would otherwise be occupied by different plants, the number of which is therefore decreasing. The one compensates for the other; the general principle remains perfectly true.

Fig. 5.5. Part of a colony of May apple. (Photo by W. H. Hodge)

The marginal plants are younger and bear mostly a single leaf, while at least those toward the center bear two leaves and produce a large white flower in May. The symmetrical shape of these colonies and their complete detachment from other similar patches indicates that each colony is the progeny, by growth of the underground stems, of a single seedling established at that spot in some past year. No one knows how long such a colony lives or what may cause its ultimate disappearance. A search of the woodlot in the spring will sometimes reveal a few young plants which, if all goes well, will eventually produce a colony, but such seedlings are rare. The propagation and perpetuation of the May apple are carried on almost exclusively by vegetative growth, but the establishment of each colony, and consequently the migration of the plant from one place to another, must depend on its reproduction by seeds.

In 1953 I visited a woodlot of an acre or two which had numerous healthy colonies of May apple, some up to ten feet in diameter. Careful search revealed but two groups of seedlings, each consisting of twenty or thirty tiny plants growing very close together and thereby showing that a May apple fruit had decayed at that spot. At the end of the year, all or nearly all of these little plants were alive, but plainly showing the effect of competition by too much crowding. The next year each clump was reduced to two plants and the third year to one plant only. There you have it; only a single seed survived!

Now consider what we have done. First, we built up the idea of the extraordinary capabilities of plants for migrating (Chapter 3). Even as we did it, we introduced the apologetic statement that most seeds come to rest near the parent plant rather than at a distance and that migration takes time, often very much time. Then we said that there are many places where a plant can not grow at all, even though it does migrate thither (Chapter 4). Lastly, we have restricted the successful migration to just one seed, on the average, from each plant. The remarkable thing is that all these ideas are true. The only one which is likely to be misinterpreted is this last one of a single successful seed from each plant. The reader must not regard it as a fixed mathematical formula but accept it as a general principle with many short-term exceptions. In fact, we shall now proceed, in the next two chapters, to see what happens if this average is not reached or is exceeded.

6/ Behavior of a Plant within Its Habitat

The Weather Bureau is able to compute the average temperature and rainfall for any day, week, or month of the year for any of its regular stations, basing its averages on the records of a long series of years. Experience shows that the same kind of weather will prevail for some distance around each station. But it is also a matter of experience that there is rarely a day, a week, or a month that does not deviate more or less from the average. A similar condition is true for each plant habitat. Within each habitat the environment, in most cases, is essentially uniform throughout the whole area at any one time, but it varies from one season to another and from one year to another with the ordinary fluctuations of climate.

Every such deviation has its effect on one or more of the numerous kinds of plants which occupy the habitat, and as a result the number and vigor of individuals vary noticeably from year to year. This is especially evident among the annual plants, which must be re-established from seeds every year. The conditions of weather which prevail during the year, the distribution of rainfall during the growing season, and fluctuation of temperature from month to month affect such plants directly and immediately, and their effect is evident to us in the size and relative number of individuals of each species. A warm week in spring may hasten the germination of seeds, and the seedlings may then be killed by an unusual late frost or a period of drought. Protracted rains at the flowering season may interfere with pollination, reduce the crop of seeds, and thereby reduce the chance for seedlings to appear the following spring.[1] In the preceding chapter, the variation in the spring display of flowers in the deserts was mentioned and attributed to variations in rainfall.

[1] I have recently seen a good example of this in a bit of dry upland woods near my home. In a little open glade, only a few feet wide, hundreds of plants of pinweed (*Lechea*) appeared in 1949. Apparently, one plant had existed there in 1948, and these were its progeny. The weather was unusually dry in 1949 and not one of these seedlings bloomed. As a result, not one plant of pinweed appeared in the glade in 1950.

Perennial herbs, already well established, are much less affected by environmental fluctuation, and the variation in their numbers is correspondingly less, but the differences of height and general vigor of growth are often conspicuous. Mature trees are almost never killed by these transitory fluctuations, but their seedlings are just as susceptible as the annual herbs. The reaction of mature trees is shown by variation in their vigor, and this is not often very apparent to the casual observer. It may be seen easily in pines and related evergreen trees. The annual growth in height of these trees is shown by a circle of lateral branches marking each year's growth. In any plantation of young pine, one may observe and confirm by actual measurement how much their growth in height has varied from year to year. Trees lay down each year a new layer of wood over the older layers in their trunk and branches. By using a special kind of auger known as an increment borer, one may obtain a slender core of wood extending from the bark to the center of the tree and can measure how the growth in thickness has varied from year to year. What is more important, these variations can be correlated directly with variations in the climate, as shown by Weather Bureau records. Having verified this correlation, we can then carry the matter further and from the growth-rings of very old trees learn something about the climate long before the Weather Bureau was established. Professor A. E. Douglass of the University of Arizona carried such observations back more than a thousand years in the southwestern states, and he demonstrated that there have been protracted periods of unusually dry or unusually moist climate there and in other parts of the western states.

During the course of a century, or perhaps less, every type of climatic fluctuation[2] has appeared in each habitat, and the plants which live there are mostly[3] the kinds which can withstand these fluctuations. The long-lived trees have certainly experienced them all. Among the short-lived species the number of individuals of each kind may vary and some species may disappear for a short time, being restored by later migration. The most remarkable example of this that I have seen was in the Middle West not long after the great drought of the nineteen-thirties. (My own observations were made in central Illinois and southwestern Iowa, but the effects were essentially similar over a wide area.) Some readers will remember that this drought lasted several years, caused widespread crop failures and economic difficulties, and

[2] This term does not include great long-term climatic changes, such as would be associated with the development of another glacial period.

[3] *Mostly* is used instead of *all* because there may be some new arrivals there now which will succumb during the first year of abnormal environment.

Fig. 6.1. Young tree of western white pine (*Pinus monticola*), showing variation in the spacing of the annual whorls of branches. (U.S. Forest Service photo by W. H. Leve)

led to the abandonment of thousands of acres of farm lands, the emigration of thousands of people, and the writing of *Grapes of Wrath.* Ordinarily a woodlot in the Middle West contains a fairly luxuriant undergrowth of perennial herbs and shrubs, but in 1940 the average woods were almost bare of these plants. No exact figures are available, but it was my general impression that the number of individuals was not more than a tenth of normal. Whether any species were completely exterminated over wide sections of the country is not known. If they were, it may be many years before the usual processes of migration will bring them back from farther east.

Most of our climatic fluctuations last only a short time, and their effect is chiefly on the short-lived kinds of plants. Also, since our own period of intelligent observation is seldom more than 50 years, we do not directly appreciate that there may also be long-term variations of climate with correspondingly slow and almost imperceptible changes in the plant life. Conversely, when we note, as we easily can, variation in the long-lived trees from one place to another, we may not realize that these variations may be caused by some local event of many years ago.

In northern Michigan I have examined carefully a large number of samples of the northern hardwood forest. It consists of several kinds of large trees, but sugar maple and beech are always the most abundant. Hemlock is usually third, but forests can be found with no hemlock at all. The herbs and shrubs which grow on the floor of these forests include many species of plants, with only few individuals of each. They vary but little from one place to another and lead one to believe that the present environment is essentially uniform in all the samples. Then why should one forest contain 48 percent sugar maple, 32 percent beech, 14 percent hemlock, and 6 percent all other trees, while a second sample contains 82 percent sugar maple, 16 percent beech, no hemlock, and only 2 percent of all other kinds? It may be that there is some quirk in the environment not apparent to direct observation which gives the maple a great advantage and is very prejudicial to hemlock in this second sample. But it is also possible, and this is the more plausible explanation, that the difference in the trees today is due to some difference of environment many years ago, which for a period of years gave the seedling maples an advantage over young beech and hemlock. One of these forests was especially interesting. It contained maple, beech, and hemlock as usual. The ground was carpeted with seedlings of maple and beech, and there were many tall saplings of both trees waiting to take a place in the forest when the present mature trees died. Hemlock seedlings were very few; of hemlock saplings there were none.

Any one would at once predict that at some future time, when the existing mature hemlocks die of old age, the forest would consist of only maple and beech. (This prophecy would have been wrong, for the whole forest was cut for lumber the next winter.) After it was cut, an examination of the growth-rings on the hemlock stumps showed that all of them had been about the same age, three hundred years or a little over. They had all been seedlings and saplings together in the first quarter of the seventeenth century. Then what happened to the environment, beginning about 1625 and continuing ever since, to prevent the development of young hemlocks in that place? And when, if ever, might the environment again become favorable to the establishment of young hemlocks? Who knows?

For all kinds of plants, there are certain times and certain places at which life is difficult or impossible. Then the plants fail to mature the one seedling necessary to replace the parent plants, and they are represented by decreasing numbers of individuals. There are also times and places in which life becomes easy and the favored plants mature more than one seedling each, and the number of individuals increases accordingly. Because we as observers are limited in our time, the scope of our observations, and the scope of our understanding, we may fail to appreciate that such conditions exist unless the changes are so rapid and evident in a small area as to be obvious.

The chickweed (*Stellaria media*) in our gardens springs up in great numbers in earliest spring but mostly disappears with the hot days of June. That is perfectly apparent to us, and we can correlate its disappearance with hot weather. But suppose that our lives were seventy hours long instead of seventy years. We would then conclude from direct observation that chickweed always had lived and always would live in our gardens, and our grandfathers of last week would confirm our conclusions. It is all a matter of time.

We sometimes read of "young and aggressive" or "old and decadent" kinds of plants. We are apt to compare them with a young man making a place for himself in the world through youthful energy and ingenuity and an old man sitting all day by the fireside and dreaming over his past. All such statements should be regarded very skeptically. A so-called "old and decadent" plant is merely one which is having trouble, at some particular time and place, in producing the one new plant to replace each old one and thereby maintain the usual number of individuals. An "aggressive" one is a plant which at some particular time and place is able to mature more than the one necessary seedling, or which finds conditions favorable for vegetative multiplication, and is accordingly increasing in number. Our chickweed is an "aggressive" plant in March and a "decadent" one in June. Of course some species are older than others, but an old species may be "aggressive" and a young one may be "decadent," or the same species may be "aggressive" in one area and "decadent" in another, depending entirely on local conditions.

The adjustment between a plant and its environment is often very delicately balanced, and even a slight change in the surroundings may cause a conspicuous change in the behavior of the plant. The showy lady slipper (*Cypripedium reginae*, Fig. 6.2), perhaps our most beautiful wild orchid, might be considered one of these "decadent" species. It is a shy plant and a rare one, growing far back in the bogs and usually only in small numbers. Yet a friend of mine, who had a very few plants in his bog, merely cut out a few trees, letting a little light into the bog, and in a few years had a fine display of the lady slippers. Clearly it is not a decadent species; it merely suffers from an oppressive environment and is quick and aggressive to take advantage of any better opportunity to grow. The giant sequoia is certainly an old species and restricted in its range to a small part of California, but it is not decadent. Countless seedlings spring up near the parent trees, and some of them, if left alone, will eventually take the place of the present generation of mature trees. *Metasequoia* (Fig. 6.3) is another very old plant. In prehistoric times it had a very extensive range over most of the North Temperate Zone, but it gradually disappeared and was regarded as extinct. It was known to us only from the numerous and often well preserved fossils. Recently it was discovered as a living plant in western China. It is not decadent. It produces good seeds which germinate well. The seedlings grow well and young trees may now be seen in many parks and private collections in this country, and it has recently been reported to have survived an Alaskan winter. It is being sold commercially by some nurseries in the United States.

There are, of course, thousands of plants known to us only by their fossilized remains. Although they are extinct, we can be sure that they were never decadent. They did not die of senility; they merely succumbed to changes of environment, worsted in the struggle for existence. One might just as well accuse Custer's army of decadence and senility.

Among individual animals, especially man, there may be a form of degeneration which can be called decadence, but it seems to be moral and ethical rather than physical in nature. Such a decadence, if it affects enough people, may have far-reaching effects; certainly it has been cited as a cause for the fall of the Roman Empire, and there are those who firmly believe that it is the greatest potential danger to the United States. Senility (as opposed to

Fig. 6.2. Showy lady-slipper (*Cypripedium reginae*) in Itasca County, Minnesota. There are five species of lady slipper in the deciduous forest region of the northeastern United States. (Photo by H. E. Stork, from National Audubon Society)

decadence) among man and other animals is shown by the degeneration or malfunction of organs, either causing death directly or leaving the individual an easier prey to bacterial and virus diseases. This sort of degeneration is much less significant in plants, which are continually producing new supplies of young leaves, twigs, and wood. The death of plants is more often due to starvation (as in competition for light and water) or to parasitic diseases, especially those caused by fungi. One eminent naturalist called attention to the lack of diseases on the giant sequoia and said that the tall existing trees might live forever if it were not for soil erosion and lightning.

In every community there are certain species of plants which are generally considered to be rare and others which are abundant. Botanists have given little attention to the causes of these conditions; indeed, they have scarcely been adequately defined.

Rarity and abundance are not precisely contrasting terms of opposite meaning. A rare species is rather to be contrasted with a common one, while abundance should be contrasted with sparsity. Rarity and commonness refer to the distribution of a species; sparsity and abundance to the number of its individuals. A rare plant is one which lives in only a small fraction of the habitats in which it might reasonably be expected; a common plant is found in most or all of them. A sparse kind is represented by a small number of individuals; an abundant one by great numbers of them.

For examples, we might turn to the pine barrens of southern New Jersey. There are scores of bogs of all shapes and sizes in this region, and the pitcher plant (*Sarracenia purpurea*, Fig. 11.3) occurs in nearly all of them; it is a common species. The bog asphodel (*Narthecium americanum*), on the contrary, occurs in only a few; it is a rare species. In any bog the pitcher plant is usually represented by a comparatively few and widely scattered individuals; it is not particularly abundant. The bog asphodel, if present at all, may be represented by thousands of individuals, making great patches of yellow at the blooming season; it is an abundant plant. These pine-barren bogs are surrounded by many square miles of dry upland sands largely covered by pine and oak. Under these trees grows another interesting plant of the region, the broom crowberry (*Corema Conradii*). One may easily spend long hours hunting for it without success, since it is both rare and sparse. In contrast with it, we may consider the leatherleaf (Fig. 4.3) of our northern bogs, present in almost every one of them and with vast numbers of individuals forming a continuous mass of shrubbery. It is both common and abundant.

No one knows much about the causes of rarity and commonness, of

Fig. 6.3. A group of metasequoias at the New York Botanical Garden. Like the closely related bald cypress of the southeastern United States, this tree drops its young twigs in the autumn and is bare of leaves in the winter. (New York Botanical Garden photo by Walter Singer)

sparsity and abundance. We may theorize about it as much as we wish, but we shall have difficulty in substantiating any of our ideas. That is because we seldom know in detail how a plant lives, what it actually requires of its environment. Small differences which may escape our notice entirely may be of critical importance to the plant. Here is one idea for your meditation: The rarity or commonness of a plant reflects (at least in part) the degree to which its migrations have been completed; the sparsity or abundance of a plant is correlated with the general suitability of the environment and its success up to the present time in its competition with other kinds of plants.

We have been trying to present a very sketchy and fragmentary idea of how a plant lives within its own habitat. It is not primarily a geographical subject. It belongs rather to the domain of plant ecology, and in a book dealing with that field it would be necessary to pursue the subject much further and to discuss various other important aspects of it. If the reader wishes to cogitate about it, he will find plenty of questions to occupy the most agile mind. In considering them and searching for the answers, he will do well to keep in mind and adapt to plant life a good old proverb: possession is nine points of the law; and one well-known saying attributed to General Forrest of the Confederate Army: success in battle depends on getting there "fustest with the mostest men." An equally important item, for the other side of the scales, is that superior force or competitive ability may in the long run compensate for relative immobility.

The behavior of a plant within its habitat has been considered here because some understanding of it will help the reader to appreciate the circumstances under which a plant migrates to new habitats. That is a geographical matter, which will be considered in the next following chapters.

7/ The Speed of Migration

In general, we have very little information on the rate of migration of plants. It is easy to see that the *seeds* of many plants travel at a rapid rate, especially those which are carried by wind, birds, or water currents, but migration depends not on the speed of travel alone but equally as much on the ability of the seeds to grow and establish new plants in the new location, and that is a very different matter.

The fact has already been emphasized that the establishment of a plant after the movement of its seeds depends on finding a favorable environment. It appears that most (not all) plants travel so fast that they have already occupied all the areas of favorable environment on their own continent, and they can not move farther until there have been changes in the environment of lands beyond their present range. Or in other words, the movement of plants is so rapid, so efficient, and so thorough, that any significant change in the environment is immediately followed by the appearance of a new set of species.

What do we mean by *immediately*, as used in the preceding sentence? When we are dealing with processes and movements which require a long time for their completion, we certainly can not make the word mean a matter of hours or minutes. It means instead that the potential migration of most (not all) plants is at least as rapid as the change of environment which enables the plants to grow in their new location.. Some environmental changes are rapid, as the denuding of a mountainside by an avalanche in a few seconds. Obviously the plants which will eventually reoccupy this area can not arrive in a similarly short time. Also, it has just been stated that not all plants do migrate so rapidly as to keep pace with environmental changes. Whether they do or not is a difficult matter to prove. It is certainly theoretically possible that they do not, and there are some instances of plant ranges which may be explained thereby. Notice the discussion of the range of the bald cypress in Chapter 9. Most environmental changes are slow, some of them very slow, as measured in terms of human activity and experience, but they are accompanied step by step by the arrival of the immigrating plants.

Speed is a function of distance and time. Most disciplines of science have suitable standards of distance. The astronomer measures in light-years, the engineer in inches or feet, the bacteriologist in micromillimeters, and the physicist in angstrom units (one of which is about four billionths of an inch long), but all of them use the familiar units of time. The geologist and the phytogeographer have no unusual difficulties in measuring distance, but they lack a standard for time. Both of them deal with periods so long that such a term as a year has little significance, and the only clock they have is the speed of deterioration of radioactive isotopes, such as carbon 14. Both of them are forced to express time in relative terms, with time in comparison to other phenomena, and that is the best we can do here. All we can say is that the speed of migration for most plants is as rapid as the change in environment.[1]

The migration of our common weeds is of quite a different character. Such plants live in areas where man has made great changes in the environment, clearing off the natural plant covering, plowing the soil, and draining the land. Weeds travel fast because they accompany man in all his activities. Their seeds are found as impurities among the seeds of crop plants and are sowed with them; they travel in hay, in ballast, in packing boxes, in cattle cars, and in various other ways often unknown and unsuspected. As a result, they arrive in a new country about as soon as the settlers. When the botanist William Baldwin visited the Middle West in 1819, he was surprised to see how every boat-landing along the Mississippi and Missouri Rivers was already populated by European weeds.

The entrance of such foreign species, followed by their rapid migration in this country, is still in progress. *Galinsoga ciliata*, one of the commonest weeds of gardens in the eastern states, is of sufficiently recent arrival that it has not yet acquired an English name. It is a native of northern South America. Being of tropical origin, it is easily killed by frost, but its seeds are frost-hardy nonetheless. It first appeared in this country, so far as known, between a century and a century and a half ago along the Gulf coast. It was probably introduced independently at other times and places as well, and in the hundred

[1] The importance of the time factor in the migration of plants is so obvious that it is surprising how little attention has been given to it by ecologists and phytogeographers. It has been most gratifying to the author to read the article on the distribution of weeds by Lindsay (*Ecology*, XXXIV [1953], 308–21), who says, "The question of time enters into all problems of plant distribution. . . . It can be concluded, then, that most if not all weeds used in this study have had time to migrate to their full geographical limits in the state."

Fig. 7.1. Purple loosestrife. (New York Botanical Garden photo)

plus years it has spread over the eastern states and northward into Canada. *Erucastrum gallicum*, a member of the mustard family, also still without a common name, was unknown in the northeastern states until a few decades ago; at least it was not mentioned in Gray's Manual of 1908 or the Illustrated Flora of 1913. It is now common in the vicinity of the Great Lakes and is rapidly spreading farther east and south. The most recent editions of Gray's Manual (1950) and the New Britton and Brown Illustrated Flora (1952) describe many other plants not appearing in the earlier editions.

The reason for the rapid migration of such plants is easily explained. These plants did not have to wait for changes in the environment before they could move. The proper environment was here and waiting for them; they needed only to arrive. Then they spread rapidly, depending less on their normal means of migration and more on the involuntary assistance of man.

Comparatively few foreign species have been able to migrate rapidly over the country by their own normal means of dispersal. The flowering rush (*Butomus umbellatus*), a rather handsome, pink-flowered swamp plant related to our common arrowheads (*Sagittaria*), is one of them. It is a native of Europe and first appeared a few years ago along the shores of the St. Lawrence River in Quebec, presumably from a shipment of pulpwood. Thence it spread, probably by its small seeds adhering to the feet of birds, up and down the shores of the St. Lawrence and later to the shores of the Great Lakes and Lake Champlain. During the next few decades it may be expected to consolidate its now widely scattered stations and also to migrate farther. The purple loosestrife (*Lythrum Salicaria*, Fig. 7.1), with long spikes of gaudy flowers in August, also prefers wet soil and has spread very rapidly during the last thirty years, probably by the same means as the flowering rush, until in many parts of the northeastern states there is scarcely a pond, lake margin, low swale, or even roadside ditch without it.

These plants which have just been mentioned are all foreigners, which on their arrival in this country found just the environment they needed and spread rapidly as a result. What they wanted was land where the natural vegetation had been removed or seriously disturbed by man, and the assistance of man in their migration. There is no indication that these plants ever traveled so fast in their native homes. Can our American plants migrate with similar speed? We have many native weeds which have spread widely across the land within historical times, such as the tumbleweed (*Amaranthus albus*) and the black-eyed Susan (*Rudbeckia hirta*), but both of them have had the advantage of human commerce to aid their migration. Excluding them and considering only the

Fig. 7.2. A large mesquite bush (*Prosopis velutina*) in southern Arizona. There are several closely related species of *Prosopis* in northern Mexico and southwestern United States. For our purposes they may all be conveniently grouped under the general name mesquite. (U.S. Forest Service photo by R. L. Hensel)

native plants of natural habitats, which must migrate by their own natural methods, we can not find any except the southwestern mesquite (*Prosopis*, Fig. 7.2) which is or has been migrating at a comparable rate within historic time. Even here it is probably a combination of overgrazing and protection from fire, both due to human influence, which has permitted the species to spread so rapidly. And yet we have hundreds of native species just as well fitted for rapid migration as the flowering rush or the purple loosestrife. Some of them, as the waterweed (*Elodea canadensis*) and the horseweed (*Conyza canadensis*), have appeared as weeds in Europe and have spread there very rapidly.

From a financial standpoint, one of the most remarkable instances of rapid migration is that of the cactus (a species of *Opuntia*) in Australia. Originally introduced from America for planting in the warmer parts of the country, it escaped from cultivation and spread at an alarming rate over the sheep country. Every year thousands of acres were added to its range and became absolutely useless for the grazing of sheep. Value of the land decreased greatly and the sheepmen lost heavily. After a considerable search, an effective insect pest of the cactus was imported into Australia, which not only stopped the spread of the weed but actually killed it. Recently a sheepman from Australia told me, "I could have bought thirty thousand acres of cactus land for ten shillings an acre but I would not take it for a gift. Now it is worth ten pounds an acre."

Comparing the behavior of American weeds abroad with that of foreign weeds here at home, we can only conclude that the migration of our plants is no slower than those of other lands, but that our plants had already migrated to their environmental limits before botanical study began. We may imagine the range of a plant as bounded by an environmental fence beyond which the plant cannot pass but which can be relocated. And we can summarize this discussion and close this chapter by repeating a sentence from its third paragraph: The potential migration of most plants is at least as rapid as the change of environment which enables them to grow in a new location.

8/ Retreating Migration

Animals, which move as individuals, may easily advance or retreat. Man may migrate into the "dust bowl" area and move out again when the climate does not suit him. Plants, rooted to the ground, have no such ability as individuals; but, as *kinds* of plants, they may appear by migration in an area not previously occupied by them and may also disappear from it. The latter may conveniently be called retreating migration, although the processes involved in it are quite different from those of a forward migration.

By this time we have learned that plants have no control over the direction in which their seeds travel. If they did have, and if all the seeds produced by plants near the edge of their range could be directed backward, then the species would disappear there with the death of the present generation. Such a movement would be very similar to the advancing migration by which plants extend their ranges. The chief difference between forward and retreating migration would be that plants have the advantage, when moving forward, of long jumps by accidental or unusual means while, when moving backward, the location of the rear guard would always be determined by the shorter movements of normal seed dispersal.

We also know that a plant soon comes to occupy all the area of favorable environment which is accessible to it, and that it can not extend its range farther unless the environment beyond the present range becomes favorable also. Conversely, a kind of plant remains in the area as long as the environment remains favorable. The individuals eventually die of old age or other causes, but the normal reproduction perpetuates the species indefinitely.

If for any reason the environment begins to worsen, the plant is at once in difficulty. Not only does each individual suffer by reduced vigor and possibly also in reduced seed production but the reproduction of the plant by seedlings is also affected. We have learned that, as a theoretical average, each individual need produce only one successful seedling to maintain the kind in normal numbers. As soon as the average drops to less than one, due to deterioration of the environment, the number of mature plants begins a gradual decrease which, if continued, eventually results in the disappearance of the plant from the area.

Now the rate of worsening of the environment is usually just as slow as the rate of improvement. That is, it is usually too slow to be evident to us from one year to the next. The fluctuations in the weather from one year to another are generally large enough to mask any progressive change in climate for a long time. In exceptional cases the rate of worsening of the environment may be rapid, as when a landslide on a mountain destroys a forest in a few seconds. It is often fast enough for a good observer to perceive the effect during his lifetime. It may be so slow that no person can detect or appreciate it, even with the help of proper instruments. It may be compared to the movements of the three hands on a watch dial. One spins so rapidly that its motion is easily visible. The minute hand appears to be stationary, but its actual motion can be detected by watching it half a minute or so, whereas the hour hand appears to be fixed in one spot. And yet it is the most important of the three in indicating the passage of time, just as the slowest environmental changes are likely to be the most important, in the long run, in controlling plant migration and therefore plant distribution. Regardless of the rate of deterioration, the effect is always the same. On the individual plant it causes a reduction of health and vigor, varying from unobservably slight to immediately fatal. On the reproduction of the plant, it causes a higher death rate of young seedlings, varying from small to a complete mortality and consequent failure to reproduce. Regardless of the rate of change, the species sooner or later disappears if the environment does not again improve.

Such is the cause and the process of retreating migrations; they have occurred repeatedly in all parts of our land. They are occurring now, and it is fairly easy to demonstrate the fact for those which are proceeding at a relatively rapid rate or affecting only a small bit of land, as we shall presently do. Indeed, if we accept the idea that every natural area is already producing plants to the limit of its capacity, then any forward migration of a plant requires a reduction in numbers or the actual disappearance of another to make room for it.

The best documented retreating migration of a species is that of the American chestnut (*Castanea dentata*, Fig. 8.1). Down to the beginning of the present century, it was a very abundant timber tree in the forests of the eastern states, extending through the mountains and the Piedmont from southern Maine to central Alabama, with outlying colonies in Michigan and Illinois. Then it was

Fig. 8.1. American chestnut (*Castanea dentata*). Left, a tree in Ohio in 1930 (U.S. Forest Service photo by L. Kellogg); right, leaves and flowers (U.S. Forest Service photo by E. R. Mosher)

attacked by a parasitic fungus which had just arrived from Japan. That long jump was one of those accidental or unusual steps, since it apparently arrived on a shipment of nursery stock of the Oriental chestnut. After it once got here, its further movements were by its normal methods, which appear to be by spores adhering to the feet of birds. The first attack on the American chestnut was in New York City. From this center, the disease spread several miles each year in an ever-widening circle, killing chestnut trees by the thousands. Today the American chestnut is almost extinct. Sprouts still come up from some of the old roots which had not yet decayed and, since they are already *old* plants, sometimes bear a few nuts. Each successive year brings fewer such sprouts, and we may expect that the American chestnut will soon be gone forever.

Possibly you may feel that this is not a migration but a battle to the death in which the chestnut invariably loses. But that is just what all retreating migrations are, although few of them are as rapid, or as widely extended, or so important economically, as that of the chestnut. Is the word *migration* a misnomer when used for such a process? Possibly it is, but it is nevertheless a very convenient word to use, and we shall continue to use it. We shall, of course, remember that an advancing migration requires actual travel of seeds (or of other reproductive bodies, such as the spores of the chestnut blight) into new territory, while a retreating migration is brought about by the death and lack of reproduction of an existing population.

9/ Limits of Range and Migration

Are our native plants still extending their range, and what sets a final limit to migration? These are difficult questions, and they can not always be answered even by the most careful observation. It is usually fairly easy for an ecologist to examine the border of a habitat and to determine whether a plant is extending its range at that point. But at the outer margin of the distributional area of a species it is usually very difficult to decide whether the species is advancing or retreating or standing still. If one travels westward along the fortieth parallel of latitude, approximately from Philadelphia to Columbus to Kansas City, he will observe that beech (*Fagus grandifolia*, Fig. 9.1) is a common and abundant tree in every habitat favorable to it, avoiding swamps and dry rocky soil and choosing the deeper and more fertile soils. This will continue as far west as Indiana and just over the state line into Illinois, and from that point west no beech will be seen. Many other kinds of forest trees continue west to and even beyond Kansas City. Is beech migrating westward but merely lagging behind its fellow trees, or retreating eastward, or is its margin stationary? Actually we do not know, nor can we answer similar questions for the vast majority of other species. By search we can find their present limits in every direction, but we can seldom get any very good clue to their further movements.

To be sure, one can get clear evidence in many places along the northern parts of the eastern hardwood forest that beech and sugar maple are extending their area at the expense of the pine and spruce forests of the north. But it is not clear whether they are actually extending the total range of the species farther north or merely expanding their local habitats. Griggs found excellent evidence in Alaska that the northern forests were actually moving out into the treeless arctic tundra, but he was not sure whether the movement was the temporary effect of a cycle of comparatively warm years or a general and permanent movement. Ramaley and Dodds found that the western yellow pine (*Pinus ponderosa*, also called ponderosa pine, Fig. 9.2) was moving out from the foothills of the Rocky Mountains onto the grassy plains of Colorado, but they also learned that this was a temporary movement caused by a cycle of years with rainfall above the average. Bessey, many years ago, indicated that the trees of the eastern forests were migrating westward along the rivers of Nebraska.

The fact remains, as stated in the preceding chapter, that the establishment of a plant in a new locality generally depends on an environmental change, and that the means of migration are so efficient that plants are usually right up to the limit of their environment and can not move further until it changes. At the utmost limit of their ranges, plants may be waiting for changes in climate, and these are always very slow.

Reverting to Bessey's generally accepted statement that trees are still migrating westward across Nebraska, what shall we say about beech, which has only reached Illinois, while hackberry, ash, and elm are already five hundred miles ahead of it? Has beech lagged behind merely because its heavy nuts travel more slowly than the winged fruits of ash or elm or the edible fruits of hackberry? If you answer yes, we shall then ask you to explain why some kinds of oak, with acorns bigger than beech nuts, have already crossed the Missouri River on their way west. Or has beech reached a point where the climate prevents further migration westward? Frankly, we do not know.

We do believe, however, that most native plants have reached the limit to which they can travel under present conditions of climate (that is, temperature and rainfall), and that further migration depends on future climatic changes. Such changes have certainly occurred in the past, will undoubtedly occur in the future, and are very likely happening right now, but they are so slow that they produce little or no demonstrable effect on the migrational behavior of any species and must be judged more by geological evidence or by a consideration of a large number of species collectively.

If we assume, then, as a matter of general belief, that many or most species have already migrated as far as they can under present conditions, what particular feature of the environment has set this limit?

The environment of a plant is very complex; it includes all the conditions, processes, and substances which affect or control its life and health. Plants require certain mineral substances which are derived from the soil; the chemical character of the soil is therefore a part of their environment. They require water which is absorbed from the soil; the water content of the soil is another factor. Plants transpire water from their leaves, and the rate of loss depends on atmospheric humidity, temperature, wind, and light. They require direct

Fig. 9.1. American beech in North Carolina. This is a relatively young tree, which will develop a thicker trunk as the years go by, although it may not get much taller. The smooth gray bark is one of the characteristic features of the species. (U.S. Forest Service photo by Paul S. Carter)

Fig. 9.2. Veteran tree of ponderosa pine in Colorado. This tree is about 110 feet tall, rather large for the area, although taller forms occur in the Pacific states. (U.S. Forest Service photo by H. D. Cochran)

or indirect sunlight for the manufacture of their carbohydrate food. They are often dependent to a greater or lesser extent on micro-organisms in the soil. They are affected, often seriously, by animals which eat them or by fungi which are parasites on them. All these factors, except the last two, are subject to modification by other plants which are growing nearby, so that one of the most important environmental factors is competition with other plants for space, water, and light. This is especially serious with seedlings, which can be overshadowed and choked out by the more rapid growth of plants already established.

The matter is further complicated by the fact that the environment necessary to a plant at one stage of its life may be different from that required at another stage. Many seeds, for example, will not germinate until they have been frozen, while the plant itself will not grow unless the temperature is well above the freezing point. That is a simple matter, and in our latitudes it is easily taken care of by the alternation of winters for freezing and summers for growing, but it would effectively prevent the growth of many northern species in the constantly warm tropics. While the demands of a plant vary, the environment itself, right in the same spot, varies even more. The available water in the soil becomes gradually less after every rain until replenished by the next. Light varies from dawn to noon and again to dusk and with every cloudy day and passing cloud. Temperature, like an elevator, is always having its ups and downs. The important point is that each plant must find at the spot where it is growing, and despite all these fluctuations, the environment that it needs or can endure. Otherwise death will result.

This whole subject is a part of ecology proper rather than phytogeography. It is introduced into these pages in several places so that it will be emphasized in the minds of our readers and they will appreciate its relation to plant distribution. If we do not know what the environmental demands of a plant are, we certainly can not decide accurately what sets a limit to its range. We can see what the problem is, we can guess at the answer, but we can not prove it.

Another complication is the fact that the various factors of the environment are to some extent compensatory. A relatively poor condition in one respect may be tolerated by the plant if other conditions are near the optimum. This relationship is well illustrated by the cultivated plants in our gardens. We plant, for example, tulips from the hot semi-desert of Persia, thyme from the Mediterranean region with its mild moist winters and long dry summers, pachysandra from the mild moist climate of Japan, and Iceland poppies from the much colder climate of northern Europe, and all of them will grow in close proximity to each other. But we compensate for the unusual conditions of temperature and rainfall to which they are subjected. We give each plant the kind of soil, the amount of water, the kind of fertilizer, the winter protection that we have learned from experience will be most beneficial to it. Above all, we keep them free from the competition of weeds and other plants.

Analogous conditions also exist for our wild plants, and we find them growing and thriving in one part of their range under conditions which they would not tolerate in another part. For example, plants which demand considerable water may be found in a variety of habitats in a region of abundant rainfall, but only in wet soil along streams in a drier climate. Thus, in the northwestern part of its range the sycamore (*Platanus occidentalis*, Fig. 9.3) is almost confined to alluvial land along streams, while in the southeastern states, where the climate is considerably moister, it grows freely on uplands as well. There are many species of herbs which are calciphiles (lime-loving) towards the margins of their range, but are relatively indifferent to the concentration of calcium in the soil in the center of their range. Others become calciphobes, avoiding limestone, as the limits of range are approached. In general, when we examine how and where plants grow at the margin of their range, we find that they occupy only some restricted habitat, while toward the center they may and often do occur in a variety of places. In this restricted marginal habitat, however, the number and vigor of the individuals may compare favorably with what we find at the center of the range.

If we compare the environmental conditions in this marginal habitat with those prevailing in other habitats near by, we may arrive at an opinion on the cause of the existing limit to the range of the species. I say opinion, not conclusion, since our evidence will be largely circumstantial and may easily lead us astray. For beech, as mentioned above, we find that it grows in a wide variety of conditions in the eastern states; in Ohio and Indiana it occupies large areas but chiefly in the moister and more fertile soils, while at the limit of its range in Illinois it is confined to hillsides, somewhat below the general level of the adjacent uplands, where the soil probably receives more water by surface runoff and by seepage from above, and where the crown of the tree does not extend as high as that of the forest above. From these observations we may form the opinion, although it may be wrong, that it is the water relations which stop the migration of the beech at this point, especially in their effect on the seedling stages, but whether it is the water supply in the soil which is primarily responsible, or the rate of loss of water from the leaves, or both together, we can only guess.

Fig. 9.3. Sycamore tree in North Carolina. This specimen is doing well on an upland site. (U.S. Forest Service photo)

Fig. 9.4. Leaves and fruits of sycamore. Sycamore leaves resemble maple leaves, but can easily be distinguished by the fact that the main veins diverge a little above the base of the blade, instead of exactly at the base as in maples. (U.S. Forest Service photo by W. D. Brush)

Let us go farther west and look at the distribution of various forest trees in northwestern Iowa. All the uplands there were originally covered with prairie. Trees grow only in the river valleys where they find a subterranean supply of water, or they extend up the narrow ravines which are eroded along every valley. In these ravines the trees grow to various heights but, in general, do not push their crowns much above the level of the adjacent uplands. Here we immediately reach the opinion that it is the exposure to wind above this level, causing a great increase in the amount of water lost by transpiration, which restricts the trees to this height and prevents them from growing out in the open uplands.

Trees extend north approximately to the Arctic Circle, in some places crossing it, in others falling somewhat short of it. Here the limiting factor naturally seems to be temperature, but this conclusion needs to be examined more closely. It is not so much the absolute minimum temperature, which occurs during the dormant season of the trees and consequently has relatively little effect on them, but rather the length of the period during which a favorable temperature continues in summer. A tree must have temperatures above the freezing point to grow; and, if it is to lay down new wood, produce dormant buds for next year's growth, and ripen seeds for future generations, it must have a reasonable temperature for a reasonable length of time. The time is as important as the degree. The required duration and the necessary temperature have often been estimated by botanists, and it seems that the temperature must average 50 degrees or more for at least eight weeks, if forest trees are to grow.

The cause of the timberline (Fig. 9.5) on mountains, above which trees do not occur, has also often been considered by botanists. As in the Arctic, the general location of the timberline is obviously a matter of temperature, but it is greatly influenced by the local distribution of snow, which is often heavy on the higher mountains. Not only must there be a summer of reasonable warmth long enough for the trees to complete their annual process of ripening seeds, producing new wood, and forming buds, but also the ground beneath them must be free of snow long enough for seeds to germinate and for seedlings to get established.

If one rides on any of the north-south railways or highways of the eastern states in winter, he can easily see the evergreen plants of mistletoe (*Phoradendron flavescens*, Fig. 9.6) on some of the trees. He can observe that mistletoe extends north as far as southern Illinois, southern Ohio, and Delaware, and that south of these limits the plant is commoner but no more abundant. Mistletoe takes its water and minerals from the plant to which it is attached, and depends on the air for its light, carbon dioxide, and oxygen. It grows on a variety of trees, and those on which it grows at its northern limit extend considerably farther north. What sets the northern limit for mistletoe? It can not be the kind of tree it grows on, since they also grow farther north; and of course the plant is independent of soil conditions and probably equally uninfluenced by rainfall. There is in fact no significant change in rainfall for a good many miles north of the plant's geographical limit. The immediate answer to the question is obviously temperature, but we do not know whether it is the length of the growing season or the minimum temperature in winter. We strongly suspect the latter, and Dr. Schneck reported many years ago that a series of severe winters had killed the mistletoe over a belt sixty miles wide.

The Spanish moss (*Tillandsia usneoides*, Fig. 1.1), which so conspicuously drapes the trees of Florida and the Gulf Coast, is not a real moss but a flowering plant. It has small greenish yellow flowers which are followed by a crop of plumed seeds, giving the plant a fine means of migration. It extends north along the Atlantic Coast as far as Virginia and southern Maryland (Worcester Co.),[1] and the farther toward the north the less it penetrates inland. It does not go nearly so far north in the Mississippi Valley. Since it lives on many kinds of trees, takes all its water and raw materials from the rain and air and dust, and has a remarkable resistance to drought, we naturally suspect that its northern limit is fixed by the minimum winter temperature, but we do not know. Spanish moss persisted on a tree near the Hudson River just north of New York City during a few successive mild winters, but we do not know whether it migrated there or was hung there by some tourist returning from Florida. Some years ago unprecedented cold weather sent the thermometer down to −18°F. at Richmond and nearly as low at Norfolk. We ought to know how Spanish moss fared that winter around the Dismal Swamp (just south of Norfolk).

It has been said, probably without conclusive supporting evidence, that the climate of Jamestown, Virginia, when the first settlement was made there in 1607, was similar to the present climate at Plymouth, Massachusetts, where the first settlers in 1620 suffered so many hardships. It is a known fact that the winters of the North Atlantic states have been distinctly warmer in recent years than the average of the last half of the nineteenth century. If the temperatures in winter have actually risen, not for any single year or small series

[1] The southern Maryland station was apparently only temporary, and the plant is no longer to be found at the spot where it seemed to be thriving only a few years ago.

Fig. 9.5. Timber line on Cover Mountain, Utah. The trees are Engelmann's spruce (*Picea Engelmannii*) and limber pine (*Pinus flexilis*). Dense, stunted growth of the sort here shown is known as Krummholz. (U.S. Forest Service photo by Lincoln Ellison)

of years, but generally during the past three centuries, there should be some discoverable effect on the distribution of plants. Unfortunately there were no botanists at Jamestown in 1607. Still, the northern *known* limit of Spanish moss was until recently the Dismal Swamp. Only a few years ago it was detected farther north across the straits in peninsular Virginia, and still later fifty miles farther and over the state line into Maryland. But does this mean a recent migration or was the plant merely overlooked by the few botanists who live in that region?

The bald cypress (*Taxodium distichum*, Fig. 9.7) is one of the more important trees of the southeastern states. It extends north in the Mississippi Valley to southwestern Indiana and southeastern Illinois, where it lives in the deep alluvial swamps along the Wabash River. The northernmost known station for it is just a trifle north of the latitude of St. Louis. It is a fact that one of the great glacial advances of many thousands of years ago extended south to just about this same latitude and must have set an absolute limit to the northern range of plants in general at that time. The alluvial soil which forms the valley land of the Wabash is washed down from farther north in glaciated territory; there is no significant difference in it, whether south or north of the glacial boundary. Also, the bald cypress, when planted by man, flourishes much farther north, not only in wet soil but also in ordinary upland soil. No one knows why it has not migrated farther north. This may be an example of a plant which migrates very slowly, has not kept up with its environment, and has not yet reached the climatic limit of its migration.

One of the most conspicuous plant boundaries in the world scarcely concerns the United States. That is the frost line, south of which (in the Northern Hemisphere) freezing temperatures never occur. Thousands of kinds of tropical plants are killed by the slightest frost, and we are probably safe in assuming that almost all lowland plants which grow only south of the frost line are comparably susceptible. We say lowland plants because on high tropical mountains there may be plenty of snow and freezing weather. The frost line crosses Florida near the southern tip of the state, where many distinctly tropical plants are native. Of course, several to many years may elapse in other parts of Florida between severe cold waves bringing freezing temperatures; and, for short or long periods, central and even northern Florida may be frost-free. Many tropical plants, therefore, extend well north of the frost line, their numbers progressively decreasing northward. When a cold wave comes, as it did in 1957-58, such plants may be killed back to the ground and reappear as sprouts from the roots, or they may survive only in very sheltered

Fig. 9.6. Mistletoe (*Phoradendron*) on a black walnut tree in Texas. (New York Botanical Garden photo)

places, or they may pass the cold period in the seed stage, but in general their northern margin is pretty well determined by the occurrence of cold weather.

There are many trees which can stand some frost but die if the temperature goes low enough and stays low long enough. Low temperature is not always the direct cause of such winter-killing; much of it is, in fact, due to physiological drought—the plant's inability to absorb enough water from the frozen soil to make up for what evaporates from its twigs and branches. A cold spell accompanied by drying winds when the ground is bare of snow is much more serious than a similar degree of cold with ample snow-cover and without wind. The giant sequoia (Fig. 1.2) withstands severe cold in its native home in the Sierra Nevada of California, where there is a deep mantle of winter snow and an ample supply of ground water. In New York state, where there may be prolonged cold spells and frozen ground without snow, the tree sooner or later succumbs. Peaches are an important commercial crop near the shores of the Great Salt Lake in Utah. A few miles away in neighboring Cache Valley, where it is just a little cooler, young peach trees may do well for several years, only to die during the first unusually severe winter.

Note that in all of the examples in the previous paragraphs it is the *unusual* winter which does the plants in. In all climatic factors affecting plant growth, it is likely to be the unusual extreme rather than the average condition, which sets the limits of range. Conversely, it may be only the unusual year with exceptionally favorable conditions which permits the establishment of seedlings of a species, yet this occasional good year may be enough to permit the perpetuation of trees and other long-lived plants.

The mesquite (*Prosopis*, Fig. 9.8) is a shrub or small tree of our southwestern deserts and formerly extended east as far as central Texas. Suppose we had tried sixty or eighty years ago to decide what set its northern and eastern boundaries. Just by observation in the field, we might have concluded that it could not migrate north into the Great Basin because of cold winters, or on the eastern side of the Rocky Mountains because of cold winters and the competition with sod-forming grasses, or straight east because of the moist climate. In recent years mesquite has begun a migration which has taken it north into Kansas and east, strange to say, across southern Louisiana and actually across the Mississippi River. We do not know or guess how much farther it may spread in another half century. It now seems that its former boundary in northern Texas may have been determined by prairie fires, and its eastern line by its intolerance of shade in forested eastern Texas and Louisiana. Prairie fires have virtually ceased, forests have been cleared, and mesquite is on the march.

Fig. 9.7. Bald cypress in Mississippi. Bald cypress is called bald because the small twigs fall off in the autumn, leaving the tree leafless in the winter. (U.S. Forest Service photo by Robert W. Neelands)

Fig. 9.8. Increase in mesquite over a period of 38 years, due mainly to a combination of overgrazing and protection from fire. Left picture taken in southern Arizona in 1903, the right in 1941 at the same spot. (U.S. Forest Service photos)

Travelers along the highways of southern Texas will pass through great groves of it, growing on land which used to be open grazing country.

In summary of this discussion we must admit that we actually *know* very little of the causes and conditions which set a limit to migration or to the range of plants. We may often form an opinion as the result of direct observation, and this opinion may well be correct, but actual proof would require very careful experimental studies of the environmental requirements of each species and equally careful measurement of the environment as it exists along the margin of the range. But it is still fun to wonder and guess about.

10/ Looking Backward on Migration

So far, we have considered only what plants are doing now and what they may do by migration in the future. We have concluded that they have the means of going forward into new territory and that they may be doing so even now, or that they have reached their limit, for the present at least, and are remaining stationary, or that they may be compelled to retreat from their present stand, either now or in the future. But we have not looked back to see where they came from.

This may be interesting. It certainly is interesting and important in human history to trace the migrations of people back through the centuries to their source, or so far back that there is no further evidence of their source. If the reader is of English ancestry, he may know that his forebears migrated to America two or three hundred years ago. If he can trace his lineage further, he will probably find that some of his ancestors migrated to England from France eight or nine hundred years ago and that their ancestors had come to France from Norway two or three hundred years before that, while other progenitors had come to England from Germany still four hundred years earlier. But whence came the Norse and the Saxons? Here historical records fail. There are words in Latin and Greek, and even in Sanskrit much like the corresponding words in Norse and German, and these indicate that all these races may have come from the same source, which we like to place somewhere in central Asia.

Suppose we assume, for the sake of argument, that the original home of all these races was in central Asia. If we consider where they migrated from this ancestral home, we shall learn something which we can use when we look into the migration of plants. These primitive people could walk, and paddle in boats, and may have had domestic animals to ride. Theoretically they could have gone in any direction they pleased and as far as they wished. But did they?

To the north of them lay the icy tundras of Siberia. They did not like that kind of climate and kept away from it. Even to this day their descendants avoid it, unless attracted by the mines of Alaska or the fisheries of Antarctica, and even then comparatively few establish permanent homes there. To the east lay the plains of China, but they did not go there, possibly because they were afraid of the Chinese or kept out by force of arms. South lay the tropics of Asia. Some of the people did cross over or through the mountains to the plains of India, but they went no farther, and even to this day white men tend to avoid the tropics. Most of the people turned west from their home in Asia and occupied all Europe until they met the impassable barrier of the Atlantic Ocean. From Europe they again turned south and occupied parts of the deserts but did not cross them into tropical Africa.[1] Much later, man learned how to cross the oceans and the tropics and then in a very short time, comparatively speaking, white men occupied South Africa, Australasia, most of America, and the temperate parts of South America.

Now the migration of plants has proceeded in exactly the same way. Every species has progressed in every direction until it met a barrier it could not cross or reached an environment in which it could not grow or thrive.

Ancient man could migrate as an individual, and modern man has developed extraordinary means of travel. There is no limit now to the distances he can go or the places he can reach, at least on earth. Plants,[2] on the other hand, are still dependent on their natural means of migration, just as were prehistoric men. Every step in the migration of a plant is limited to the distance which its seeds can travel between ripening and germination. Draw a circle around your maple tree which will mark the farthest distance to which its seeds are blown by the wind. Every young maple of the next generation is limited to that circle which is determined by the location of the parent tree. This parent, too, was once a seedling, and the seed which produced it came from an older tree also located somewhere within a similar distance.

[1] The migrational history of white men here suggested may or may not be substantially correct. Something like it has commonly been postulated among anthropologists, but a more recent evaluation of the evidence puts less stress on relatively recent migration and more on evolution *in situ*. It has even been suggested that the major races of man were differentiated before they reached the evolutionary level at which they could be called Homo sapiens, and that these several regional geographic races of Homo erectus (the ancestor of the Homo sapiens) evolved more or less concurrently into modern man. Such a mass evolution of one species into another is believed to have occurred among certain kinds of plants as well as animals.

[2] Excluding, of course, weeds and plants of economic or horticultural value which are transported accidentally or purposely by man.

From these considerations we draw one important general conclusion: The present distribution of every species of plant is determined not merely by the present environment but also by the location of its ancestors in the past. This also is an axiomatic statement, but, like many other statements of equal simplicity, it seems to have been neglected by plant geographers. The noted German botanist Adolf Engler was apparently the first to appreciate its importance in plant geography and to express it in print about 1880. A little thought on the matter is worth twenty pages of discussion. In one's imagination, one may trace back to ancestors of a plant as many thousands of years and as many hundreds of miles as he wishes.

In plant geography we are not interested in the ancestry of an individual plant; we are concerned rather with kinds of plants. If that maple mentioned above happens to be a sugar maple, we know that today many thousands of them are growing along the northern limit of the species, through southern Canada all the way from eastern Quebec to Manitoba. We want to know whether these trees as a group, as a species, are just holding their own, or advancing farther north, or gradually retreating. And we also want to look back and find where their ancestors were living, not a single generation ago, but a thousand, or ten thousand, or a hundred thousand years ago. If they were not living in southern Canada, we want to know why, and why they were able to migrate there later.

These are some of the real problems in phytogeography. It may seem that such a problem can not be solved. There are, however, various sorts of evidence which we can find and which, if properly pieced together, can lead us to really plausible conclusions. We shall come to some of them later.

Fig. 11.1. Goldenrod (*Solidago canadensis*). This is the common form of *Solidago canadensis* in most of the eastern United States; it has often been treated as a distinct species, *Solidago altissima.* (Photo by Molly Adams, from National Audubon Society)

11/ Looking Backward Still Further

As long as we accept the general conclusion reached in the preceding chapter—and it is obviously true—we shall try to look further and further back in the ancestry of a plant to discover its original home. But is there no limit to this question?

There is. We reach the limit with the origin of the species.

Species have not existed forever. We can dig up the fossil remains of thousands of kinds of plants and find among them scarcely any which are living today. Clearly those which are living now have come into existence at some more recent time. The younger the rocks from which we excavate the fossils, the more numerous are the species which still exist. This shows that they did not all originate at the same time. There has been a gradual appearance of species. Some of those living today are already very old; some are very young. For most existing species we do not know the relative age, and all we can say is that we are confident that they were not existent very far back in geological time.

When a species comes into being, its individuals must have some means of reproduction and usually some method of seed dispersal. They must either use these and establish successive generations, or else the species becomes extinct at once. Probably many species do fail right at their origin; we have no interest here in such plants. Nor are we concerned with the method by which new kinds of plants originate, except as the nature of their origin is related to geographic distribution. We must digress into the field of evolution just far enough to consider this point.

How can a new species come into being? How can the special characteristics, or combination of characteristics, which mark a species originate? One way is by the crossing of two different species,[1] producing progeny

[1] Although many readers probably understand the meaning of *crossing* and the process of seed production in plants, a brief explanation may be desirable here. In ordinary plants there is

unlike either of the parents. Such new kinds of plants are called *hybrids*, and the process is *hybridization*. But that merely pushes the question one step further back. How did the parents of the hybrid acquire *their* distinguishing features? The answer is clear that in general these differences result from spontaneous chemical changes in the microscopic bodies, called *chromosomes*, which are present in every living cell of the individual and which are passed on from one generation to the next as the bearers of heredity. Each chromosome has many individual hereditary determiners called *genes*, and it is these genes which occasionally undergo a persistent change which is passed on to successive generations. Such a change is called a *mutation*, and the different individuals resulting from a mutation are called *mutants*. Less commonly, there may be more gross changes in whole chromosomes or parts of chromosomes, or changes in the number of chromosomes. Changes of this sort may also, in a broad sense, be called mutations.

Mutants may be quite obviously different from the original form; those which are obviously different are often called *sports*, in the nursery trade. Many of our cultivated varieties of apples and other fruits originated as sports, and are propagated vegetatively by grafting. More often, a mutant is only slightly different from its progenitor; oftener still, the difference is so slight as to be masked by environmental differences.

We do not know all the reasons why plants mutate and produce mutants. Cosmic radiation is thought to be one cause. Extreme or unusual environmental conditions of various sorts are another, although the resulting mutants are not noticeably better adapted to these extremes. But plants continue to mutate even when grown under simulated normal conditions in a chamber protected from cosmic and other mutagenic radiation by lead shielding. Doubtless some mutations are due merely to the inherent chemical instability of organic matter.

One thing, however, is certain: The same type of mutant may appear

an organ (or several of them) within the flower, called the pistil. The lower part of it, the ovary, contains one to many minute structures, the ovules, which may later become seeds. Plants also produce organs called stamens, commonly in the same flowers as the pistils, but often only in different flowers. These produce a great many microscopic or dustlike pollen grains which are carried to the pistil, sometimes by honey-gathering insects, sometimes by wind, sometimes by other means. Once arrived on the pistil, a pollen grain begins to grow into a long, slender, but still microscopic structure, the pollen tube, which traverses the tissues of the pistil and penetrates the ovule. There further complicated processes take place, as a result of which the ovule develops into a seed, while the ovary matures into a fruit. By *crossing* is meant the use of pollen of one kind of plant to start the development of seed in a different kind. Almost any textbook of general botany will provide a more detailed discussion of this whole process.

repeatedly. That is, many different individuals of the same species may, at different times and places, produce exactly the same mutant. This fact appears very clearly in genetic studies of the little fruit fly Drosophila. These flies lay their eggs on fruit, and the larvae flourish on a diet of bananas in a quart milk bottle. When the adult insects appear a few days later, the whole brood can be examined and the mutants discovered. Much of our knowledge of mutation is based on such studies. In practically every genetics laboratory these flies are grown to demonstrate the principles of genetics and evolution, and the same kinds of mutants appear repeatedly. The same phenomenon can be observed in plants, but not so conspicuously, since plants can not readily be grown in such great numbers or in such a short time. Even more remarkable than the repeated appearance of the same mutants is the strange fact that many of them tend to appear at an approximately uniform rate for each kind.

The same phenomenon undoubtedly exists in nature among our wild plants. Perhaps the most obvious example is the production of albino flowers. Almost every kind of plant with blue, purple, or red flowers may occasionally produce white ones. These albinos appear widely scattered in space and time, and may attract our attention if the flowers are reasonably large and conspicuous.[2]

Mutants, such as we have been discussing, are not new species in themselves; most of them are not even potential new species. They are ordinarily interfertile with the parental types, and they merely increase the variability of the species to which they belong. Practically all species show some genetic variability as a result of mutation, and the hereditary differences among the individuals of a species are the result of past mutation.

There are several ways in which hereditary variability within a species can be translated into the origin of a new species; all involve differential reproduction or survival among the genetically differing individuals of the species. There may be a gradual change in the whole population of the species, through generation after generation of natural selection among the numerous seedlings, until after a great many generations the cumulative change warrants the treatment of the later population as a species distinct from the earlier one. Some botanists think that the Pleistocene fossil species *Acer torontoniensis* gave rise to our present sugar maple (*Acer saccharum*) in this way. Or the species may spread over such a large geographic area that the differing environmental

[2] One of the commonest of these albino mutants is the white-flowered violets which occasionally appear in a patch of the normal blue-flowered plants. One of the most conspicuous is a white-flowered pickerelweed in a swamp full of blue ones.

conditions in different parts of the range lead to its divergence, through natural selection, into two or more species, each adapted to its own region. The geographic varieties which we see in many species may be halfway steps in this process, and indeed botanists often disagree as to how such geographically differentiated groups should be treated. One of our common goldenrods (*Solidago canadensis* Fig. 11.1) is treated in the New Britton and Brown Illustrated Flora (1952) as being a transcontinental species with several regional varieties, whereas in the current edition of Gray's Manual (1950) some of these regional populations are treated as distinct species.

Probably the commonest of all the methods by which new species arise from old involves relatively rapid changes in an isolated fragment of the population, where the conditions of selection may be rather different from that over most of the range of the species. In a very small population, even the accidents of survival or death become important in determining the characteristics of the group. These potential new species are not at all uncommon, but most of them either succumb to the vicissitudes of competition or are swamped out by interbreeding with the original parent type when some change in the circumstances permits the restoration of geographic continuity with it. It is only the exceptional one among these many temporary local variants which goes on to become a full-fledged new species.

Now let us return to the origin of species by hybridization. The occurrence of hybrids in nature is beyond doubt. In some groups they are observed fairly frequently, in others rarely or never. The geneticist working in his experimental garden can make hybrids between so many different kinds of plants that it is easy to get an exaggerated idea of the importance of hybridization in nature. All he needs to do is pick up a little pollen on a camel's-hair brush and daub it on the pistil of another kind. Maybe it will be successful; maybe it will not, but at least he can try. If the flowers of the two species are not in bloom at the same time, he can often preserve the pollen several days in the refrigerator.

There are many factors which restrict the natural hybridization of plants. The parent plants must bloom at the same time, so that the pollen of one is available when the pistil of the other is ready to receive it. This makes hybridization easier between plants with a long season of bloom. They must be closely related to each other; only rarely can plants of different genera be crossed successfully,[3] and then they must be of the same botanical family. If they are

[3] Common in our florists' shops are some beautiful orchids developed by the artificial crossing of species of *Laelia* with species of *Cattleya*, and collectively known as *Laeliocattleya*.

Fig. 11.2. Black-eyed Susan in Maine. (U.S. National Park Service photo)

pollinated by insects (and most plants are), the same insect must visit both parents within a short span of time. Insects do not usually cover much ground in their flights, and most of them, especially many species of bees, prefer to specialize each day on one kind of plant.[4] Plants can obviously be farther apart if they are pollinated by wind, which may carry the pollen for many miles.

If cross-pollination has been effected and viable seed produced, there are still difficulties. Some hybrids are weak and unable to grow to maturity. Some are unable to reproduce (the domestic mule is a familiar example). Some hybrids are so like one or both of their parents that they back-cross with them and thereby disappear. Still others can not reproduce themselves and their seeds produce plants more nearly resembling the parents instead.[5] If crossing is to be fully successful and produce a permanent new kind of plant, the hybrid (1) must be healthy and be able to grow to maturity, (2) must be able to produce plants like itself, (3) must be so unlike either parent that it may properly be called a new kind, and (4a) must not back-cross readily with either parent or (4b) must migrate so far from its parents that it is physically unable to back-cross with them.

Distinct species have been created by hybridization in experimental work. A few species which exist wild in nature have been re-created by artificial crossing, as the hemp nettle (*Galeopsis tetrahit*), and we can conclude that they arose originally in the same way. A new species of cord grass (*Spartina*) originated in England during this century as a result of hybridization between the native English species and a species introduced from America. Many other species in nature are believed to have arisen by hybridization, although

[4] Dr. A. B. Stout, noted geneticist, grew in his garden plants of the cardinal flower (*Lobelia Cardinalis*) and the great blue lobelia (*Lobelia siphilitica*) and found that they hybridized easily when artificially cross-pollinated. That raised the question of why natural hybrids are so seldom seen, since the two species often grow close together in wet soil and bloom at the same time. Dr. Stout then grew plants very close together, brought their flower-clusters into actual contact, and tied them there. Some flowers of the cardinal flower were removed and flowers of the great blue lobelia allowed to remain in place of them. The result was a big spike bearing red flowers at the bottom and blue at the top. Then he watched the behavior of the visiting bees (bumblebees). They usually began work at the bottom of the cluster of red flowers and gradually progressed upward. Pretty soon, and apparently to their great surprise and confusion, they came to blue flowers which they would not visit. They buzzed around angrily, often tried it again, but soon flew away. In this instance, bumblebees would not visit different species of flowers even when it would have been easier to do so than not. No wonder hybridization takes place so seldom in nature.

[5] This statement is true enough for our present purpose; the careful student may wish to read further in a genetics textbook.

we do not yet have experimental proof. The origin of new species by hybridization may or may not involve also a change in the number of chromosomes in each cell of the individual; for our purpose here, these differences are immaterial.

Now if individuals of species A can cross naturally with species B and thereby produce species C, they may do it repeatedly and they may do it at any place where plants of A and B are growing together. The origin of species C may be at any time and place within the overlapping ranges of the two parent species. The original distribution of a species of hybrid origin is therefore not necessarily a point but may be several points or an area, an area which may possibly be extensive. We have already noted that species which arise by natural selection without hybridization may also differentiate over a wide area rather than at a single point. It is only the species which originate as small local populations which consistently have a single time and place of origin, and although this may be the commonest method by which species originate, it is by no means the only one. Furthermore, since we do not know how most of the particular species now existing did arise, the possibility of regional rather than point origin must generally be considered.

This opens a new vista in plant geography. If we try to locate the ancestors of a plant and have no evidence to help us except the present distribution of the species, we are very soon baffled. Let us find then some very closely similar plant, which we infer (almost certainly correctly) is actually related to our plant. We do not know whether one has evolved from the other or both have evolved from a third species, but in either case the geographical distribution of both species depends on their migration from the area of origin, and by knowing this distribution we may get some clue as to the location of their origin. For example, the familiar black-eyed Susan (*Rudbeckia hirta*, Fig. 11.2) is one of several species of *Rudbeckia*, all native to North America. We know that black-eyed Susan has migrated far and wide in the United States in recent times, but the location of its relatives leads us to conclude that its origin was somewhere in the eastern half of our country. On the other hand, this kind of evidence may not be available or, if used alone, may be of little value. It is most useful when we have a pair of closely related species, and we shall come to that matter shortly.

Some geographical features of evolution are still to be considered. A new species can rarely succeed in the same habitat and geographic area as the established parent species. A potential new species which begins its existence as an isolated fragment of the parent species depends on the continuance of that isolation for its own evolutionary divergence and separate establishment. If the connection to the parent species is restored before the divergence is complete, the potential new species is likely be to reabsorbed into the parent population. The same is true of hybrids. Since the parents of the hybrid necessarily grow close together, the hybrid offspring probably grows close to both of them. Usually, a hybrid that sets any seed at all can back-cross with either parent. Since such hybrids may appear repeatedly, the characters of each parent species are by back-crossing gradually introduced into the other. If the hybridization and back-crossing are sufficiently extensive, what were two distinct species may be blended in the course of time into a single, more variable species.

We began this chapter by stating that new species often appear as a result either of hybridization or of rapid divergence in small populations. In the preceding paragraph, we presented reasons for believing that these hybrids and local populations disappear into the parent species, producing merely variable old species instead of new ones. How do we reconcile these apparently conflicting statements?

First, some hybrids and peculiar types of mutants (involving change in chromosome number) are able to reproduce themselves and cannot back-cross readily with either parent. These plants are safe from being reabsorbed. They can develop into a large population and migrate far and wide, provided they are able to do so in spite of strenous competition with other plants. *Galax aphylla*, commonly marketed for Christmas greenery, is a good example. It is a native of the southern Appalachian Mountains and has evolved into two chromosome-races, not sufficiently unlike to call two distinct species, but so unlike that they do not cross with each other. These two races live more or less associated with each other essentially throughout the range of the species, and each maintains its own character.

Second, suppose that a hybrid or other potential new species is adapted to a new type of environment in which its parents do not grow. Remember that hybrids and small local populations appear repeatedly and that seeds migrate into all sorts of places and sometimes to unexpected distances. Suppose that sooner or later the seeds of such a potential new species do migrate into this new type of environment to which they are suited. There they can grow, free from the danger of back-crossing with their parents, can establish a population of their own kind, and, in due time, migrate as far as this type of environment extends. Then we shall have two different species, closely related to each other, one of them, in fact, arisen from the other, and occupying

Fig. 11.3. Pitcher plant (*Sarracenia purpurea*). This species is resistant to cold and extends northward as far as Labrador. (Photo by W. H. Hodge)

Fig. 11.4. Pitcher plant (*Sarracenia flava*). This species is sensitive to cold (although it can stand some frost) and extends northward only to southeastern Virginia. (Photo by W. H. Hodge)

two different geographical areas or two different habitats within the same area.

This state of affairs is so well illustrated among the higher animals that David Starr Jordan, eminent zoologist and first president of Stanford University, long ago proclaimed the general principle which has often been called Jordan's Law: A pair of closely related species do not live together, but in different and adjacent geographic regions or in different habitats in the same region. This law is equally well illustrated among plants, and we need cite only a few examples. In the eastern states there are two species of narrow-leaved phlox (*Phlox subulata*), well-known to gardeners as moss pink or rock pink, and *Phlox bifida*. They are undoubtedly closely related to each other, having arisen from a common ancestor and not very long ago, geologically speaking. They have the same general habit, appearance, and structure, but differ in a number of small details. Hybrids between them are known in cultivation. *Phlox subulata* occupies, in general terms, the Appalachian area, while to the west of it *Phlox bifida* lives in sandy soils of the Mississippi Valley and around the Great Lakes. Since in nature they never grow close enough together to hybridize, they are able to maintain their specific identities. The painted trillium (*Trillium undulatum*) is a familiar species of the northeastern states, growing from eastern Quebec to Manitoba and thence southward to Michigan, Pennsylvania, and northern New Jersey. This area forms, roughly, a great inverted triangle and is reminiscent of the range of the pitcher plant, to be discussed later in this chapter. The nearest relative of the painted trillium is the dwarf white trillium (*Trillium nivale*), which grows from western Pennsylvania to southern Minnesota and Missouri, a range which adjoins that of its painted cousin, but scarcely overlaps it. Every botanist who studies the relationship and evolutionary origin of species meets with many other similarly related pairs of species.

If we know something about the origin of a species, what do we know about its final disappearance? All species eventually become extinct, as least so we believe, but the processes of evolution are always at work and the lost species are steadily replaced by new ones. If we consider any one particular kind of plant, we seldom know what caused its extinction. Paleontologists often wonder about the abrupt disappearance of the whole race of dinosaurs, but they have no generally accepted theory of explanation. In the plant world, we have one very remarkable case in recent years—not quite complete to be sure, for a fraction of one percent of the plants still live, and that is the American chestnut (*Castanea dentata*, Fig. 8.1). In the discussion of retreating migrations, we stated that this catastrophe was caused by a parasitic fungus disease. It may well be that other kinds of plants, and animals also, have been exterminated by disease, but we have no proof of it.

Geographically, it seems fairly obvious that the disappearance of a species is the culmination of a retreating migration. Plants commonly keep up with environmental changes and occupy all the area in which the environment is favorable. Therefore, if the environment begins to worsen, they may have no place to which they can retreat. There is a gradual change in the delicate balance between death rate and birth rate—the former increasing, the latter decreasing—until in the course of time the whole species is gone.

Environmental changes of this sort are often unilateral, starting at one margin of the range and gradually progressing toward the other. The retreating migration accordingly begins at this exposed margin, and the last surviving colonies of the plant grow on the opposite margin of the range, or at least in some spot where the destructive changes have been the slowest. Fortunately for plants, it often happens that a change for the worse at one margin is accompanied by a change for the better at another margin. Thus if the climate of the country were to become gradually colder with the approach of a new glacial period, our northern spruces and firs would be driven out of Canada. The same cooling would affect the southern states and possibly open up additional territory where these trees could grow. So there would be a retreating migration from the north, and an advancing migration toward the south at the same time. In America, where the main mountain ranges trend more or less north and south, such migrations in accordance with the temperature are easy. In Europe, where the Pyrenees, Alps, and related mountains form a barrier with an east-west axis, many species must have been exterminated during the ice ages.

It would be very nice if we could follow the general principles presented in these first eleven chapters and give the geographical history of the various kinds of plants. We would like to begin with the time and place of their origin and state what their ancestors were. Then we would describe their various migrations, how they extended in this direction or that, how they retreated from this or that boundary, and what were the environmental changes that caused the advances and the retreats. Lastly we would explain the present limit of their range, tell what determines it, and show whether, as a result of environmental changes now in progress, the plants are right now advancing or retreating farther.

Unfortunately, we can not do it. We simply do not know enough. But for some plants we can make an attempt at it, and for such an attempt we shall choose the northeastern pitcher plant (*Sarracenia purpurea*, Fig. 11.3) sometimes also called the sidesaddle flower.

There are only about ten kinds of pitcher plants of the genus *Sarracenia* in existence, and all of them grow in the southeastern states. They all like to grow in bogs where the peat is wet and acid, and most of them prefer sandy bogs where the peat is relatively shallow. Our best guess is that the whole group originated in that area, perhaps in northern Florida or southern Georgia, but we have no clear idea what their ancestors may have been like. From this center they have spread westward along the Gulf coast to various distances and even leaped the broad delta of the Mississippi River to appear again in bogs of southern Louisiana and southeastern Texas. All of this area has mild winters, and in these bogs several successive years might pass without freezing weather. Some of these pitcher plants were more resistant to cold than others and migrated northward, always keeping on the Coastal Plain, across South Carolina and North Carolina, and two of them reached southern Virginia. One of them, *Sarracenia flava* (Fig. 11.4), found its northern limit there. The other, *Sarracenia purpurea*, more resistant to the cold, kept on moving northward. How did it migrate? Probably in mud on the feet of birds, the way so many other swamp plants have traveled. It crossed the Virginia Straits into Maryland and the Delaware Bay into New Jersey.

The northern part of New Jersey and all the lands to the north of it had been covered by ice during the glacial period. After the retreat of the ice, thousands of shallow depressions were left on the irregular surface of the newly uncovered soil. If these depressions were wide enough, water drained into the bottom of them and accumulated until the water level reached the lowest spot in the surrounding land, forming a lake or pond. There the water overflowed as a brook or river. But if the drainage area (watershed) was small, or the surrounding land a little higher, enough water never accumulated to cause an overflow. Loss of water was by seepage and evaporation, and and undrained swamp or pond was formed. Around the latter in wet acid soil, and all over the former, bog plants colonized and peat accumulated.

This acid peat was just what our pitcher plant needed. Straightway it was over the glacial boundary and off on another migration northward. No longer was it restricted to the narrow coastal plain, for these northern bogs extended far inland. It spread northwest, north, and northeast until today it occupies a great upside-down triangle, with its base reaching from Labrador to Saskatchewan, its apex in New Jersey, its eastern side along the Atlantic Ocean, and the western side a line extending across Pennsylvania and Ohio, northern Indiana, northeastern Illinois, Wisconsin, and Minnesota. And it still retains its original range as a narrow strip along the Coastal Plain from New Jersey to Louisiana. This is by far the largest range of any species in the whole pitcher plant family.

Apparently the decisive causative factor in this huge migration was resistance to cold, not merely in the seed stage, which would enable the species to survive a cold winter even if last year's plants were all killed, but a resistance of the whole plant, until now it can be frozen solid for several months during a bitterly cold subarctic winter. Possibly it has not even yet reached its limit. Possibly future millennia will show our pitcher plant migrating into the arctic tundra from Baffin Bay to the Bering Straits, and even hopping across into Siberia.

A brief discussion of the deductive reasoning used in plant geography was given in our introductory chapter. This story of the pitcher plant has been developed in this way. The discussion went on to say "if *all* the pertinent facts have been considered and if the relation between the theory [our story above] and the facts has been *correctly* interpreted, this one theory is the correct explanation." Do you suppose the author has had *all* the pertinent facts at his command? Almost certainly not. Did he interpret the known facts correctly? We hope so. At least he has reached what has been referred to previously as a "plausible theory."

In summary, we may say that the origin of a species does not always take place at a single time and place, but may occur over a considerable area and even at various times. The perpetuation of a new species and its establishment as an independent member of the plant community are hampered by many difficulties, but commonly depend on its establishment in a new area or its inability to hybridize with its immediate relatives or to back-cross with its parent species.

12/ Summary of Chapters 2-11 Introduction to the Following Chapters

In the several preceding chapters we have tried to consider and explain the geographical distribution of a single kind of plant; not one particular kind only, but any kind, since all kinds are subject to the same general laws. We have often had to present or to exemplify the discussion by referring to particular kinds. These examples have usually been chosen from the plants with which the author is most familiar. There are thousands of other kinds which would have served equally well.

Each kind of plant, we have found, has a geographical range which can be discovered by travel and search, a range which may be restricted or extensive. Always it shows discontinuous distribution. Within its range it is limited to certain habitats only, in which the environment suits it, and is absent from other habitats. Within its habitat it has to share the space with other plants and lives in constant competition with them. So rugged is this competition that the plant can, on the average, produce only one descendant which will live to full maturity, although it probably produces hundreds or thousands of offspring which succumb to competition early.

The seeds or fruits of each kind of plant have some means of migration—some of them efficient, some inefficient—so that they come to rest and possibly grow at some distance from the parent. The parent plant has no control over the distance or the direction of migration, and many of the seeds may fall in places where growth is impossible. This unfortunate fact is circumvented by the production of great numbers of seeds. Besides the normal means of migration, seeds may and often do travel by unusual or accidental methods and, in that way, may reach places much farther away from the parent plant.

By the migration of its seeds or fruits, beginning at the time of origin of the species by evolution and continuing generation after generation through long periods of time, the present range has been attained. The margin of this range usually marks the geographic limit of the environment in which the plant may successfully grow, because plants usually keep up to their environment, but it may occasionally indicate merely the present extent of a migration still in progress. In reaching its present range, the plant may have been subjected to vicissitudes as well as favorable conditions. Its range at times in the past may have been larger than at present, and a change of environment may have forced it back to a smaller area. We rarely *know* what features of the environment set a limit to the migration of a plant and thereby determine its present range, but we can often make a shrewd guess. We repeat a sentence from our first chapter: Every feature of the general distribution of plants over the world is due to the combination, in varying patterns, of the separate individual distributions of the various kinds of plants.

Putting all these general principles and conclusions together, we can understand and appreciate that the range of a plant is a variable area, an area which is constantly changing as the environment changes. Of course we do not expect any appreciable change in a year or any considerable change in a century, but through the ages the ranges of plants must have changed greatly, because we know that the environments have changed greatly. Ranges have been extended in every direction and have retreated from every direction, not once but repeatedly; they have become larger or smaller in total area; they have changed their position on the map.

All of us have seen slow-motion films, in which exposures made at very short intervals have been projected on the screen at a normal rate, so that the rapid motions of an athlete have been slowed down to a snail's pace. Many of us have seen the converse, the time-lapse pictures in which exposures are made at very long intervals and then projected at the usual rate. Then very slow processes, so slow that they are not directly visible, are shown as if they were rapid. Suppose that we could have a great series of maps showing the range of the American elm (Fig. 2.4), a series made at intervals of thirty or thirty-five years for the past million years. We photograph these on a movie film, after the manner of Disney's animated cartoons, and then project them on the screen at the rate of 16 frames per second. It would take about thirty minutes to run the film, and it would give us a very vivid exposition of the changing range of the elm. We would see it retreat from the north several times as the glaciers pushed down from northern Canada and then move back toward the north again as the glaciers disappeared. We would see it leave the west and retreat toward the east as the climate of the midlands became too dry for forest trees, and push west again as the rainfall increased. And just at the end of the film, in the last five or six seconds, we would see several long tongues push rapidly toward the west—rapidly in the picture, but these few seconds would represent three thousand years!

While the preceding chapters have been used to discuss the conditions and processes which determine the range of any individual species, they have not included definite factual information about the range of any species in particular. Such data would be rather futile, since there are probably 300,000 different kinds of flowering plants in the world and probably 20,000 of them in the United States. For factual information, it will be much better to discuss plants by groups.

This involves the matter of classification of plants into groups. Since many persons are not familiar with the fundamental principles of classification, it may be well to discuss the subject right now, so that we may proceed with some knowledge of the basic ideas.

Classification consists of the organization (either mental or physical) of units into groups. Philosophers and logicians, when they consider a subject, call the units *species* and the groups *genera*. The same terms are also used in the classification of plants, but the species and genera of the botanist are by no means the same as those of the philosopher. In the simplest stage of classification, the units are individual objects, as the books in a library, the persons in a town, or the plants in an area. The groups which are organized from and composed of these individuals are always based on similarity in some chosen respect.

Suppose, for example, that the thousands of objects in a hardware store were all mixed together in one big pile, and that you haphazardly divided the pile into half a dozen smaller ones. That would not be classification, since each smaller pile would still include a miscellaneous assortment. But suppose you threw all the various kinds of nails into one pile, all the hammers into another, all the hinges into a third, and so on. That would be a real classification, since all the items in each pile would be more or less alike.

Next you may go a step further and place into a larger pile all the comparatively small piles of screws, nails, bolts, and rivets, all used to fasten things together; all the hammers, screw drivers, and wrenches into a second pile, as tools used in connection with the first pile, and so on. These new and larger piles are also based on similarity of their component parts, a similarity based on their use, and you will observe that the hammers, which were a *group* in our first sorting or classification, have now become a *unit* of a more comprehensive group.

There are no restrictions whatever on the kind of similarity which is used as the basis of classification. Almost invariably the groups are so organized that they serve some useful purpose or (especially in science) that we can learn some useful lesson from them. In botany, one of the most useful purposes is to enable us to talk or write conveniently about a large number of plants at once. We have already had to introduce one kind of botanical group into use in this book. Since an individual plant is rooted to one spot, we have to consider the geographic distribution of a kind of plant, and that is a group composed of many individuals.

It is commonly possible and often useful to classify objects in two or more ways, using different features of similarity as a basis. In our imaginary hardware store, one might prefer to classify the hammers with the nails, the screws with the screw drivers, the bolts with the wrenches. That would also serve a useful purpose and might be used in preference in another store. Obviously both systems could not be used in one store.

This hardware store or any other store represents a physical classification, in which the objects are actually placed in the predetermined groups. So also is the arrangement of books in a library, or the beds of vegetables in a kitchen garden. In science, on the contrary, classification is mostly purely mental. The units are not physically brought together in groups, indeed they could not be; instead they are mentally referred to groups for more convenient comprehension and study. Always these groups are based on one or more features of similarity; and, again, many features of similarity may be chosen as a basis.

So much for hypothetical illustrations. In the study of plant geography there are three systems of classification which can be used to advantage, and every kind of plant can be classified under each of the three systems. The first system is based on similarity of habitats and includes plants which regularly grow together. It includes many component groups: for example, the giant cactus, the mesquite, and the ocotillo which grow together in an Arizona desert; the sugar maple, maidenhair fern, beech, and trillium which live together in an eastern forest; the redbud, buckeye, and mountain lilac which live together in the foothills of the Sierra Nevada. The second system is based on similarity of general structure and gross appearance and also includes various groups, such as deciduous trees, water plants with floating leaves, climbing vines, or plants which store water for use during periods of drought. The third system attempts to reflect evolutionary origin and includes plants which are related by descent from common ancestors, such as the palms, the oaks, or the maples. This third system, called the taxonomic system, is based on everything that we know about plants, rather than on some one or few chosen features. The Latin names used for plants by botanists are names in the taxonomic system.

No matter which of these systems is considered, we find that each group which may be distinguished has a fairly definite pattern of geographic distribution, a distribution which can be studied and found to be logical and, we trust, understandable, a distribution from which we can draw general conclusions of phytogeographical interest and importance. We also find that the groups which we can distinguish under one system can often be correlated with those of another system. We see that plants of groups which *live together* often tend to *look alike*, and that groups of plants having a *similar origin* by evolution also often tend to *look alike*. And, on the contrary, we may be surprised to find that groups related by their evolutionary origin often do not live together.

We shall discuss, in sequence, some geographical features of plants which live together, and of plants which look alike. The third system, reflecting supposed evolutionary relationships, is also of interest to the plant geographer, but our attention to it will necessarily be confined to the elucidation of some particular examples. Very few species and not many genera of flowering plants are really cosmopolitan. A number of families, such as the sunflower family, occur in nearly all parts of the world, but even these are much more abundant in some regions than in others. Many other families have more or less restricted ranges, being largely or wholly confined to tropical regions (e.g., the palm family), or to temperate and boreal regions (e.g., the pink family), or to the Old World (e.g., the epacris family), or to the New World (e.g., the phlox family), or to the northern hemisphere (e.g., the oleaster family), to the southern hemisphere (e,g., the protea family), or to some more narrowly or otherwise differently defined area. Each family, genus, and species has its own geographic distribution, which reflects the diversity within the group, its history, and the ecological amplitude of its included units. Taxa of higher rank than families have relatively little if any phytogeographic significance. Since there are more than 300 families of higher plants and more than 10,000 genera, a consideration of any large proportion of these groups in an introductory book is obviously impossible.

Fig. 13.1. Gray birch. Left, a young tree on a stripping bank in Pennsylvania. Right, leaves and female inflorescences. This species is a characteristic invader of old fields and otherwise disturbed habitats in the northeastern United States. (U.S. Forest Service photos, left by B. W. Muir, right by W. D. Brush)

13/ The Joint Migration of Species into New Habitats

The fact that there are kinds of plants which habitually live together is simply stated. Adequate evidence that this is an actual fact and not merely theory can scarcely be presented on the printed page and is best obtained by each reader for himself. Unless he lives in a region wholly occupied by agriculture, such as most of the corn belt, or the San Joaquin Valley, he will easily find in his immediate neighborhood a series of spots with obviously similar environment. It may be a series of peat bogs, wet grassy meadows, sand dunes, upland woodlots, canebrakes, rock cliffs, or any of many other habitats, depending on his location. In the separate spots of any one of these series, he will find the same kinds of plants again and again and will soon appreciate that these kinds are regularly associated. He will soon realize that the kinds of plants, not one kind alone but all kinds collectively, indicate the nature of the environment and that the general nature of the environment in each habitat largely determines the kinds of plants which occupy it. Naturally, he will examine habitats which have not been too greatly changed by man. Those which have been recently cleared or drained or heavily grazed seldom illustrate the principle well.

The cause of this common association of a certain group of species in the same kind of habitat is easily understood. All of them find in that habitat an environment in which they are able to grow and maintain themselves from year to year. They are fellow citizens of the community, and, just as men in their communities, they are also competitors. It is this communal life, this sharing of space by different kinds of plants, that leads to the first and simplest kind of discontinuous distribution.

We must not assume that all the plants thus associated in a habitat have precisely the same environment. In some habitats the environment may be the same for all plants, or nearly the same. That would be the condition on a sand dune in the east or an open desert in the west. In many habitats, however, there is a distinct difference in the environments of some of the plants. That of a forest tree, for example, with its crown of foliage exposed to full sunlight, is quite unlike that of a forest herb growing in the dense shade beneath the trees.

All the various matters which pertain to the nature and conditions of the communal life of plants belong to the field of ecology rather than plant geography. To us the one important point is that these communities do exist.

Every kind of these associated plants has its own means of migration and uses them independently of the others. If the seeds of one plant reach a certain spot and can grow successfully, the chances are that the other kinds can grow there too. If they *do* live together in one place, they probably *can* in another. So, where one kind of plant migrates; the others are likely to go also. Let us call this the parallel or joint migration of plants. Neither *parallel* nor *joint* is entirely apropos, but they are short and convenient. There is no concerted or cooperative process in the migration, as one might infer from the word *joint*. There is no geometrical similarity in space or time, as might be suggested by *parallel*. Each kind of plant moves for itself. Each one migrates into various other habitats, provided there are such within migrating distance, and fails to grow in most of them. But if one of them finds a new and favorable site where it can grow, the others will almost certainly find it also. It all depends on the production of plenty of seeds and the possession of adequate means of migration.

We must not expect that all kinds of plants will reach a new habitat at the same time or in equal numbers. The time of arrival and the number of immigrants depend on these same two conditions, the quantity of seed production and the means of migration. Some must arrive first and others later. The abandoned fields which are unfortunately abundant in New England illustrate this principle well. They were originally covered by forest. Now that they have been abandoned, the forest comes back. And what are the first trees to appear? The gray birch (*Betula populifolia*, Fig. 13.1), the aspen (*Populus tremuloides*, Fig. 13.2), and the pitch pine (*Pinus rigida*, Fig. 13.3)—all with winged seeds which are easily scattered by wind. It may be many years before all the plants of the adjacent woodlots have arrived and re-established a forest essentially like the original one.

As already stated, in many habitats the environment of some plants differs from that of others. In almost every instance, this difference is due to the behavior of the plants themselves. In a forest, it is the trees, living in sunlight, which make the shade for the bloodroots and violets beneath them; it is the leaves of the trees which decompose into the humus in which the maidenhair fern (*Adiantum pedatum*, Fig. 13.4) grows. If this community of plants has the

Fig. 13.2. Aspen tree in Colorado. The same species occurs in New England and across Canada. (U.S. Forest Service photo by E. S. Shipp)

opportunity to move into a new spot, the trees must naturally go first. The maidenhair and the bloodroot (*Sanguinaria canadensis*, Fig. 13.5) will follow in due time when the trees cast enough shade and have made enough humus.

We have been talking about plants migrating together into new habitats. Where are these new habitats? Why are they new? How do they come into being?

Get into your car, with someone else to drive so you can look, and take a ride over your own country. Notice the number of different habitats that you can see—wet land, dry land; well drained land, poorly drained; steep hills, gentle slopes, level fields; stony land, sandy land, loam. Most of them will be indicated by visibly different communities of plants. In fact, you may see different groups of plants growing in what at first seem to be identical habitats, but a closer investigation will almost always reveal significant differences.

Each of these various kinds of habitats depends on features of the land itself, and any physiographer or geologist will tell you that the land is constantly changing. This is not the place to consider in detail the causes or the nature of the changes. That would take us into the field of physiography, a knowledge of which is very important to the phytogeographer, but we must introduce a few general statements. The action of water is of first importance in the humid parts of our country. Erosion by running water deepens and widens ravines and valleys, removes soil from the uplands, changes the course of streams, drains some ponds and lakes and fills up others. Water leaches fertility from the soil and, assisted by freezing and thawing, tears down cliffs and disintegrates rock into soil. Wind moves sand, piles up sand dunes, and excavates hollows between them. In arid regions the wind etches rocks, helps to reduce them to sand, and piles up deposits of wind-blown soil. Plants themselves produce humus and change the character of the soil by their presence. Atmospheric oxygen, penetrating the upper layers of soil, changes its chemical composition. All these processes go on continuously and have been going on for ages past.

There is possibly no clearer example of environmental change and the accompanying migrations of plants than those going on constantly at thousands of the little inland lakes and ponds of our northern states. The duration of a small body of water in our climate, while usually long in terms of human experience, is short in terms of earth history. It occupies a small depression in the surface of the land. Water drains into it from all sides; if the depression is large enough, there may be brooks flowing into it. All the inflowing water carries earth which has been eroding from the surrounding slopes. The pond serves as a settling basin and this earth accumulates on the bottom, gradually

Fig. 13.3. Pitch pine in New Jersey. This is the typical pine of the New Jersey pine barrens, as well as a characteristic invader of old fields in New England. (U.S. Forest Service photo by Lee Prater)

Fig. 13.4. Maidenhair fern. The common name refers to the slender, dark brown or blackish leafstalks. (Photo by W. Bryant Tyrrell, from National Audubon Society)

filling the pond. Much of the silt is dropped close to the shore, where little deltas are formed. As it accumulates there, the pond becomes not only smaller in area but also shallower. In the pond, especially in water not more than eight or ten feet deep, water plants grow in abundance. The annual crop of leaves and stems which they produce sinks to the bottom, decays into muck, and aids in the filling process. If there is an outlet stream, the current erodes its bottom and sides and it gradually becomes deeper. The deeper it becomes, the more water it carries off and the greater becomes its erosive power. It consequently tends to drain the pond, gradually lowering the water level, while the silt and muck tend to fill it up. As a consequence of both processes, the pond eventually disappears. It is converted first into a swamp and later into low wet valley-land, possibly with a stream meandering through it.

Many kinds of plants grow in and around such a little pond. (There may be a central area with water too deep for plants other than microscopic algae.) Farthest from the shore are submersed plants, as the pondweeds (*Potamogeton*) and various other kinds, growing in water as much as eight or ten feet deep. Next in order toward the shore, where the water is shallower, are plants attached to the bottom but with floating leaves, such as the water lily (*Nymphaea*) or the water shield (*Brasenia*). Next to them on the shoreward side, growing in shallow water or muddy soil, are amphibious plants, with their stems and leaves rising into the air, as cattail (*Typha*, Fig. 13.7), bur reed (*Sparganium*), and arrowhead (*Sagittaria*). Back of them, living in damp soil, are the marsh plants, including many kinds of grasses and sedges, swamp goldenrod (*Solidago patula*), joe-pye weed (*Eupatorium maculatum*) and purple loosestrife (*Lythrum Salicaria*, Fig. 7.3). These kinds of plants are mentioned merely as familiar examples. Several to many kinds will be found growing together, and in the reader's own neighborhood the most conspicuous or most abundant plants may be different species not mentioned here.

Now if our pond were perfectly symmetrical in every respect, these four types of plant life would form concentric rings surrounding the deeper water at the middle. Actually a pond is never wholly symmetrical. The slope of the bottom varies from place to place, and the deposition of soil is more rapid on some sides than others. The rings of plants therefore vary in their width and are commonly interrupted and incomplete.

As the pond becomes shallower, the central ring now occupied by pondweeds will become shallow enough for water lilies. The pondweed will migrate farther toward the center, into places where the water was too deep, and will retreat from the landward side, where it will be replaced by water lily. The latter will retreat from the part now occupied by it and the cattail will advance into its place. Around the edge of the pond, as the shore line moves inward, the cattail will migrate toward the center and the joe-pye weed will come in. All four rings of plants gradually constrict and move toward the center of the pond. Eventually, as the pond becomes almost full of silt and muck, the pondweed will disappear and the water lily will occupy the very center. Later it too will disappear and the cattail will grow at the center. Finally all open water will be filled and the marsh grasses and joe-pye weed will occupy the whole area. Each group of plants has advanced in one direction and has retreated from the opposite direction as the habitat changed, and finally disappeared as the environment became wholly unsuitable to it.

Common also in our northern states are undrained depressions with such a small drainage area that very little silt is washed into them. Here the filling of the pond is done almost entirely by the accumulation of vegetable matter. There may be pondweeds and water lilies toward the deeper open water near the center, but shoreward from them, instead of cattail, there is usually a dense mat of sphagnum moss or bog sedge, of course with various associated species. Surrounding them comes a ring of leatherleaf (*Chamaedaphne calyculata*, Fig. 4.3) and other bog shrubs, and lastly a ring of tamarack (*Larix laricina*) and black spruce (*Picea mariana*) trees. As dead vegetable matter accumulates, it is gradually compressed, solidified, and carbonized into peat, and such bogs are commonly called peat bogs. As the bog fills with peat, the first three zones move toward the center and disappear one after the other, and the site of the bog is finally occupied by a forest of tamarack and black spruce.

A careful consideration of these and other instances of environmental change which leads to a change in the plant life enables us to draw some general conclusions. First, the change is shared by several to many species: the kinds which live and migrate together. They continue to live together in their new habitat, just as they had in the old one. Do not assume that they all get there the same day, or even the same year, because some move more slowly and the establishment of some can not occur until certain others have first appeared. Second, the change is a progressive one. It begins at the margin of the habitat, and the change is in the direction of the environment of the adjoining habitat. The plants commonly extend their range gradually, with no necessity—in fact, with no opportunity—for making long migratory jumps. Naturally, as one set of plants advances, the other set must retreat. Very often the nature of the environmental change is such that, as a set of species retreats along one margin of its habitat, it is able to advance on the other margin.

Fig. 13.5. Bloodroot. The name refers to the bright red juice which is especially abundant in the roots. (Photo by W. H. Hodge)

Fig. 13.6. American lotus lily (*Nelumbo lutea*). Both the floating leaves and the emergent leaves here belong to the same species. Like other water lilies, the lotus lily grows in quiet water and is rooted to the bottom. (New York Botanical Garden photo)

Fig. 13.7. Cattail (*Typha latifolia*) swamp. (New York Botanical Garden photo)

Fig. 13.8. Black Spruce-tamarack-muskeg swamp in Wisconsin, with a meandering stream passing through it. (U.S. Forest Service photo by R. Dale Saunders)

Fig. 13.9. Early stage in the breakdown of rock into soil by plants. The plants here are mostly lichens and mosses. (U.S. Forest Service photo by Russell K. LeBarron)

This is the case in the filling of a pond, as described above. Third, the changes in the environment may be caused wholly or partly by physiographic processes, or wholly or largely by the effect of the plants themselves in controlling or modifying the environment.

Is this progressive change in plant life, this joint migration of associated species, an actual fact, or is it just the speculation of some theorist? It is an absolute fact, and any reader in the northeastern quarter of the country can, by his own observation, demonstrate the truth of vegetational change consequent on the filling of ponds. He may even find, as the author has, an old resident who has seen these changes during his own lifetime and who will say that he used to go fishing or swimming in deep water where now is only a muddy marsh. It may have taken sixty or seventy years for that change, and this example of habitat change was deliberately chosen because of its rapidity. More rapid changes may be seen on sand dunes and on recent deposits of volcanic ash, but most processes are slower and may require thousands of years for their completion.

If the reader has reflected seriously on the preceding statements in this chapter, he may have discovered in them a condition of affairs which puzzles him. This change in the plant life around a pond obviously does not lead to real extension of the ranges of the plants concerned. In fact, instead of producing a range extension, it leads to the ultimate extinction of the plants at each separate habitat. If we mentally project these processes far enough into the future, we can envision the final filling of every lake and pond and the extinction of all pond plants. Are these fears real or imaginary?

They are not imaginary by any means, but they are exaggerated. All through the northeastern states one finds thousands of areas of level soil surrounded by higher land and with a tendency to be rather wet after rains or in a wet season. If they are small areas, too small to be worth artificial drainage, they are often neglected by the farmers and are covered with a forest of American elm (*Ulmus americana*, Fig. 13.10), silver maple (*Acer saccharinum*, Fig. 13.11), and other trees which like wet soil. Physiographers can tell by an examination of the soil in such spots that they mark the location of a pond which has passed through its cycle of development and has disappeared forever. The present little lakes and ponds which occur by the thousand in Wisconsin, Michigan, New York, and Maine, and are even more numerous in Minnesota, are all on new land, on land which has very recently, geologically speaking, been worked over by glaciers. The glaciers left behind them, not a well developed dendritic drainage system, but all of these little depressions which

Fig. 13.10. American elm. Like many of our forest trees, the American elm has small, inconspicuous flowers that open early in the spring, before the leaves are out. The plants are wind-pollinated, and the numerous stamens stand out in the open air. (U.S. Forest Service photo by W. D. Brush)

Fig. 13.11. Silver maple. This species is often planted as a shade tree, partly because of its rapid growth. Like some other fast-growing species, however, it has an affinity for sewer pipes, in which the roots may proliferate so much that they impede drainage. (U.S. Forest Service photo by W. D. Brush)

soon filled up with water. Thousands of them have already disappeared; thousands still remain to form a characteristic feature of the landscape in our northern states. Most of the remaining ones are visibly on their way out, in the manner described above. Others, such as Lake Winnipesaukee and Moosehead Lake in the east and Lake Tahoe in the west, are so large that scarcely any apparent progress has been made toward their extinction.

To the south of the glaciated area of the northern states, a much longer time has been available for the extinction of any lakes which may have existed during the early history of the area. Lakes are few and far between in all the southern states except Florida. (The numerous fish-ponds, often named as lakes or ponds in the southeastern states, are man-made in recent years by damming small streams.)

We have still left unanswered a part of the possible objection from the reader, that we have not discussed the actual expansion of the range of a species into brand new territory. This has not been forgotten but merely postponed until some other ideas have been introduced, and a discussion will be found in Chapter 17.

14/ The Establishment of Joint Ranges

The conditions determining the range of a single kind of plant have already been discussed. By this time we should all understand that migration begins at the evolutionary origin of a species and proceeds from its area of origin, that the migration continues in every direction as far as the nature of the environment will permit, and that future changes in the environment may permit wider migration or may compel a contraction of range.

It is obviously to be expected that plants which regularly live together in the same environment and which are concerned in local migratory movement, such as were described in Chapter 13, will, as a result of long-continued migration, ultimately develop about the same total range. If a plant which grows *here* can also grow *there*, it may be expected to migrate *there* in due time. If other kinds of plants grow with it *here*, they can be expected to migrate *there* also. The group of plants which grow together *here* may be expected to grow together *there*.

Experience shows that this is an actual fact but, like all conclusions in phytogeography, a fact subject to many exceptions. Nevertheless, we must not let the exceptions blind us to the general truth of the statement. We must, on the other hand, always remember that these kinds of plants which grow together *here* migrate each one for itself and by its own means, and that some must necessarily arrive *there* before others.

The principle is emphatically illustrated by the thousands of peat bogs, large and small, scattered across the northern states from Maine to Minnesota, each one separated from the next by an interval, narrow or wide. Over this interval the peat bog plants have to leap in their migrations to find the next favorable habitat. And this migration, not by one species but by many, has been so successful that in any of these bogs, whether in Minnesota or Maine, one expects to find a particular group of plants growing together: black spruce (*Picea mariana*, Fig. 14.1) and tamarack (*Larix laricina*, Fig. 14.2) for the trees;

Fig. 14.1. Black spruce trees in northern Michigan. (U.S. Forest Service photo by Rexford Starling.) Inset: Leaves and cones. (U.S. Forest Service photo by W. D. Brush)

Fig. 14.2. Tamarack or American larch trees in Minnesota. (U.S. Forest Service photo by F. H. Eyre.) Inset: Leaves and cones. (U.S. Forest Service photo by W. D. Brush)

leatherleaf (*Chamaedaphne calyculata*, Fig. 4.3), bog rosemary (*Andromeda glaucophylla*), swamp laurel (*Kalmia polifolia*), and Labrador tea (*Ledum groenlandicum*) among the shrubs; pitcher plant (*Sarracenia purpurea*, Fig. 11.3), sundew (*Drosera*), and cotton grass (*Eriophorum*, Fig. 22.5) for the herbs, and under all a mat of Sphagnum moss. These plants, through their migratory ability, have come to occupy the same general area of the country. They have established what we term a *joint range*.[1]

The reader who tries to test the truth of these statements by personal observation will need much time, since he will have to drive his car several thousand miles. For our purpose it will be better to bring together some of the well known facts.

The United States Forest Service has published a series of small maps illustrating, with considerable accuracy, the ranges of many of our forest trees. Twelve of these have been selected for use here: black oak (*Quercus velutina*), northern red oak (*Quercus borealis*, Fig. 14.3), white oak (*Quercus alba*, Fig. 18.1), shagbark hickory (*Carya ovata*, Fig. 14.4), bitternut hickory (*Carya cordiformis*), beech (*Fagus grandifolia*, Fig. 9.1), sugar maple (*Acer saccharum*, Fig. 14.5), silver maple (*Acer saccharinum*, Fig. 13.11), American elm (*Ulmus americana*, Fig. 2.4), tulip tree, or yellow poplar (*Liriodendron Tulipifera*, Fig. 14.6), black walnut (*Juglans nigra*), and white ash (*Fraxinus americana*). These have been chosen because they all grow in the northeastern quarter of the United States, where doubtless many of our readers live, and because they are so common that many of us are acquainted with all or a large fraction of them. Indeed, if a reader lives in this quadrant of the country, two-thirds or more of them are probably growing within a few miles of his home. Now study the twelve small maps (Figs. 14.7, 14.8) on which the range of these trees is shown, and notice how remarkably alike they are. To be sure, some of them extend farther in one direction than do others, but in general they are very similar, and all twelve occur together in most states of the eastern half of the country.

These dozen trees are not the only ones which show this same general range. A second set might have been chosen instead, or a third, or even a fourth. So many are not necessary to illustrate our point, and the number was limited to twelve because the maps will fit nicely on two pages. This particular set of

[1] The term *joint*, although short and very convenient, is in one respect a misnomer. In human affairs the word joint usually implies a *mutual* relation of some kind, in either cause or result. Here there is no mutuality; each kind of plant migrates independently, and the word *joint* connotes merely a coincidence of results.

Fig. 14.3. Northern red oak (*Quercus borealis*) in North Carolina. (U.S. Forest Service photo by Paul S. Carter)

Fig. 14.4. Shagbark hickory. (U.S. Forest Service photos by W. D. Brush)

Fig. 14.5. Leaves of sugar maple. (New York Botanical Garden photo)

Fig. 14.6. Tulip tree leaves and flowers. (U.S. Forest Service photo by W. D. Brush)

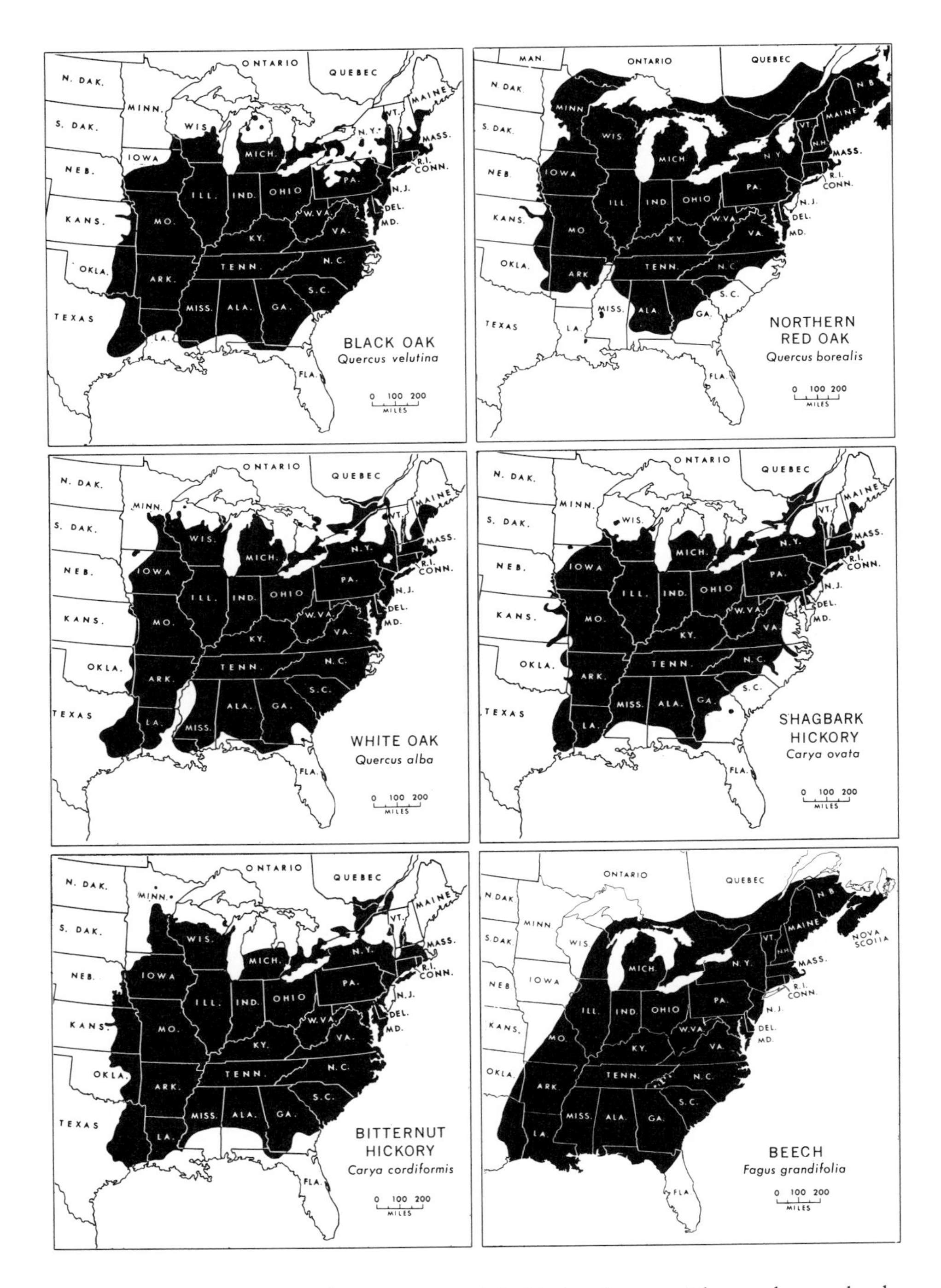

Fig. 14.7. Ranges of certain forest trees: top left, black oak; top right, northern red oak; middle left, white oak; middle right, shagbark hickory; bottom left, bitternut hickory; bottom right, beech.

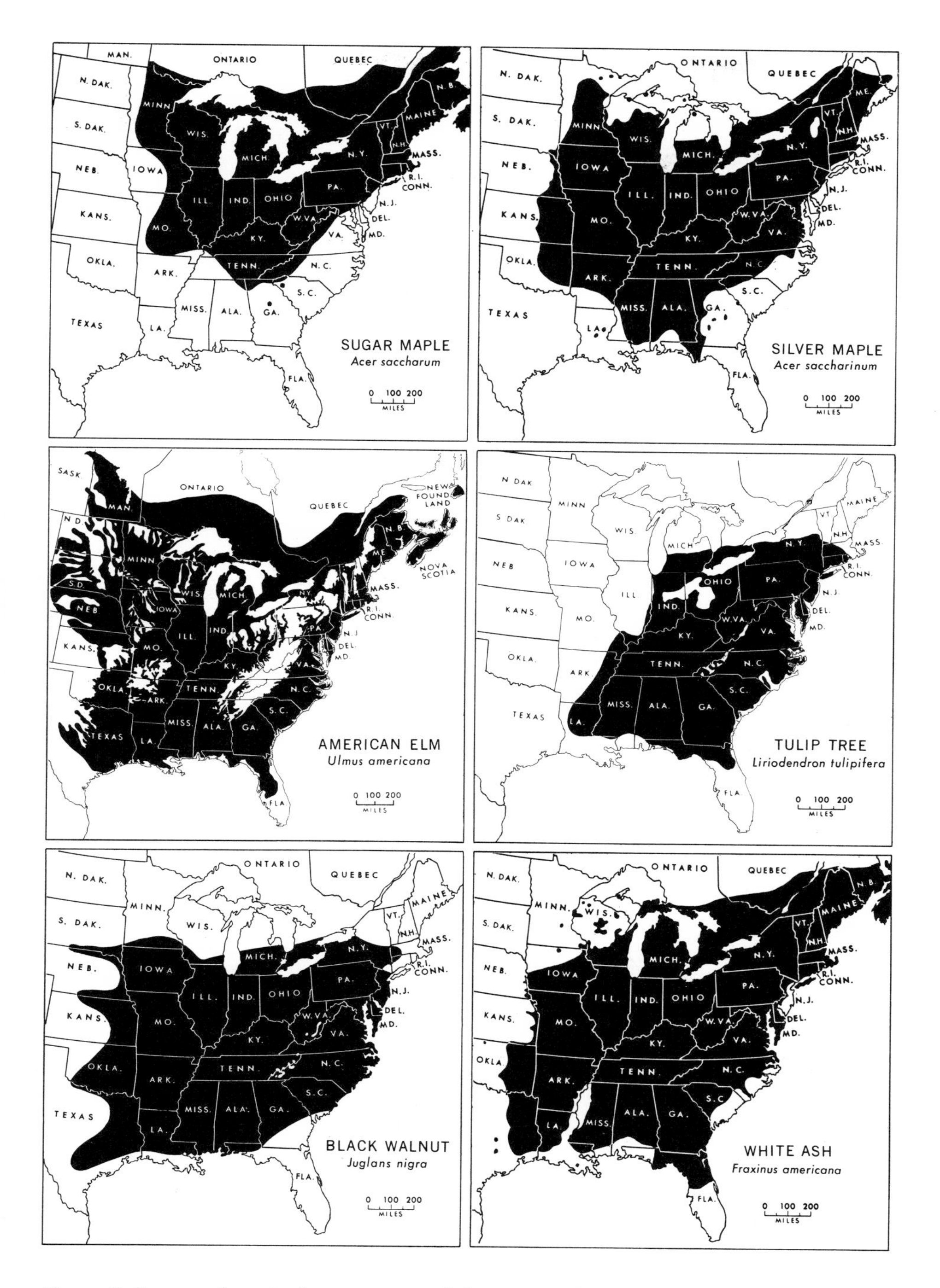

Fig. 14.8. Ranges of certain forest trees: top left, sugar maple; top right, silver maple; middle left, American elm; middle right, tulip tree; bottom left, black walnut; bottom right, white ash.

twelve was chosen partly because the maps were available and partly (and confessedly) because they do illustrate the range pretty well. Could other trees of the same region have been chosen which would not illustrate the same range? Yes, we might have chosen jack oak (*Quercus ellipsoidalis*), restricted to a small area in the Middle West, or the osage orange (*Maclura pomifera*, Fig. 3.10), restricted to the Ozarks, or the silver bell (*Halesia carolina*, Fig. 16.3), found only in the southern Appalachians. But a dozen such trees, plotted together on the same map, would again collectively develop just about the same range as shown by each of these twelve separately.

It is not necessary to restrict our illustration to trees. It is not so easy to make a map showing the range of herbaceous species, because they have not been given as much field study as the trees, and the details of their ranges are mostly not so well known. Here are fifteen well-known wild flowers which would show the same general distribution if they were plotted on a map: bloodroot (*Sanguinaria canadensis*, Fig. 13.5), Dutchman's-breeches (*Dicentra Cucullaria*, Fig. 14.9), dogtooth violet (*Erythronium americanum*), hepatica (*Hepatica americana*, Fig. 14.11), Indian turnip (*Arisaema triphyllum*, Fig. 14.12), May apple (*Podophyllum peltatum*, Fig. 5.5), meadow rue (*Thalictrum dioicum*), Sweet William (*Phlox divaricata*), toothwort (*Dentaria laciniata*), trillium (*Trillium grandiflorum*, Fig. 14.13), waterleaf (*Hydrophyllum virginianum*), wild geranium (*Geranium maculatum*), wild ginger (*Asarum canadense*, Fig. 14.14), yellow violet (*Viola pubescens*), and rue anemone (*Anemonella thalictroides*).

We shall examine the distribution of these plants by states, and we shall observe immediately that there is a large area in the eastern part of the country (and adjacent Canada) where all fifteen species exist. This area occupies Ontario (where most of the plants are found only in the southern portion), Vermont, New York, New Jersey, Pennsylvania, Delaware, Maryland, Virginia, Ohio, West Virginia, Kentucky, Michigan, Indiana, Wisconsin, Illinois, Minnesota, Iowa, and Missouri. Outside of this large area, the bordering states lack some of the fifteen. There are 14 in Quebec, mostly in the southwestern corner of the province, Tennessee, and Alabama, mostly in the northern half of the state only; 13 in Massachusetts, Rhode Island and Connecticut combined, North Carolina (mostly in the central and western parts), and Arkansas, chiefly in the northern half; there are twelve in New Hampshire, Georgia (mostly in the northern half), and Nebraska (almost all in the eastern third); 11 in South Carolina and Kansas (chiefly in the eastern third), 10 in Maine; 8 in North Dakota (eastern fourth); 7 in Louisiana (northern half) and Manitoba (southeastern part); 6 in Florida, mostly close to the

Fig. 14.9. Dutchman's-breeches (*Dicentra Cucullaria*). (New York Botanical Garden photo)

TABLE 1. Distribution of selected plant species

	Bloodroot	*Dogtooth Violet*	*Dutchman's Breeches*	*Hepatica*	*Indian Turnip*	*May Apple*	*Meadow Rue*	*Rue Anemone*	*Sweet William*	*Toothwort*	*Trillium*	*Waterleaf*	*Wild Geranium*	*Wild Ginger*	*Yellow Violet*	*Total species*
Sw. Quebec	x	x	x	x	x	x	x	x	x	x	x	x		x	x	14
Maine	x	x	x	x	x		x				x		x	x	x	10
New Hampshire	x	x	x	x	x		x	x			x	x	x	x	x	12
Massachussetts	x	x	x	x	x		x	x		x	x	x	x	x	x	13
Rhode Island	x	x	x					x		x			x		x	7
Connecticut	x	x	x	x	x		x	x		x	x	x	x	x	x	13
Vermont	x	x	x	x	x	x	x	x	x	x	x	x	x	x	x	15
S. Ontario	x	x	x	x	x	x	x	x	x	x	x	x	x	x	x	15
New York	x	x	x	x	x	x	x	x	x	x	x	x	x	x	x	15
New Jersey	x	x	x	x	x	x	x	x	x	x	x	x	x	x	x	15
Pennsylvania	x	x	x	x	x	x	x	x	x	x	x	x	x	x	x	15
Delaware	x	x	x	x	x	x	x	x	x	x	x	x	x	x	x	15
Maryland	x	x	x	x	x	x	x	x	x	x	x	x	x	x	x	15
Virginia	x	x	x	x	x	x	x	x	x	x	x	x	x	x	x	15
West Virginia	x	x	x	x	x	x	x	x	x	x	x	x	x	x	x	15
Ohio	x	x	x	x	x	x	x	x	x	x	x	x	x	x	x	15
Michigan	x	x	x	x	x	x	x	x	x	x	x	x	x	x	x	15

Indiana	x	x	x	x	x	x	x	x	x	x	x	x	x	x	x	15
Kentucky	x	x	x	x	x	x	x	x	x	x	x	x	x	x	x	15
Wisconsin	x	x	x	x	x	x	x	x	x	x	x	x	x	x	x	15
Illinois	x	x	x	x	x	x	x	x	x	x	x	x	x	x	x	15
Minnesota	x	x	x	x	x	x	x	x	x	x	x	x	x	x	x	15
Iowa	x	x	x	x	x	x	x	x	x	x	x	x	x	x	x	15
Missouri	x	x	x	x	x	x	x	x	x	x	x	x	x	x	x	15
Arkansas	x	x	x		x	x	x	x	x	x	x	x	x	x		13
North Carolina	x	x	x	x	x	x	x	x	x	x	x		x	x		13
South Carolina	x	x	x	x	x	x	x	x	x	x			x			11
Georgia	x	x	x	x	x	x	x	x	x	x	x		x			12
Tennessee	x	x		x	x	x	x	x	x	x	x	x	x	x	x	14
Alabama	x	x	x	x	x	x	x	x	x	x	x	x	x	x		14
N. Florida	x		x	x		x		x	x							6
N. Mississippi	x				x	x		x		x						5
N. Louisiana	x	x	x		x	x			x	x						7
Ne. Texas	x				x	x			x							4
Oklahoma		x			x	x			x							4
E. Kansas	x	x	x		x	x		x	x	x		x	x	x		11
E Nebraska	x	x	x		x	x		x	x	x		x	x	x	x	12
E. South Dakota	x							x					x		x	4
E. North Dakota	x		x		x		x					x	x	x	x	8
Manitoba	x			x	x	x						x	x	x		7

Fig. 14.10. Dogtooth violet. This species is *Erythronium albidum*, which has blue-white flowers and a more western, more restricted range than the closely related, yellow-flowered *Erythronium americanum*. The common name is misleading, since these plants are lilies rather than violets. (Photo by W. H. Hodge)

Georgia boundary; 5 in Mississippi; 4 in Oklahoma, South Dakota, and Texas.

We see at once that the distribution of both the herbs and the trees is about the same. There is a great area extending from Vermont to Missouri, from Minnesota to Virginia, in which all, or essentially all, of these plants live. To the north and northeast, to the south and west, the number is greatly reduced. We have picked out twenty-seven species which have coincident ranges. Having demonstrated that such coincident ranges exist, how can we explain them?

For a limited area, such as was discussed in the preceding chapter, it was stated that the segregation of groups of plants into local habitats was caused by differences in relatively small features of the local environment. Within such a limited area, as a county of average size, there is little appreciable difference in general climate. But as one goes farther and farther in any direction, the slight differences in climate accumulate until they become of great importance. Considering these twenty-seven plants, and their numerous associates also with the same general range, we naturally assume that their northern boundary is set by some feature of the temperature, and we suspect that it is probably the shorter and cooler summers which are chiefly responsible, rather than the colder winters. As we move toward the western boundary, we find that the rainfall in summer is slightly less and in winter considerably less, while the temperatures are little changed (that is, they are little changed along the same latitude). We assume that this western boundary is in some way connected with the water supply and we suspect that it is the dry air and soil of winter and the comparatively dry air of summer rather than the actual amount of rainfall. Toward the south the winters become milder and shorter, the summers longer but scarcely warmer. Many of the plants do not actually reach the Gulf Coast, and it seems probable that the boundary is set by the character of the soil as much as by the climate.

At any boundary of this group of plants, or of any other similar group of associated species in any part of the world, as the climate gradually deteriorates for the group, it gradually improves for the plants living in the next region. Our plants are therefore at a disadvantage in the competition for space. General climate, whether temperature or rainfall, usually varies gradually from place to place; soils and topography may and often do change abruptly. And as the topography changes, the local climates also show significant differences in small details, largely caused by the degree of exposure to the wind and sun. For some hundreds of miles northeast and southwest along the Atlantic Coast, the rainfall scarcely varies from the usual 40 to 50 inches per year. For a thousand miles north from the Gulf Coast to Lake Superior, the average temperature of the growing season changes by only about 20 degrees, or about one degree to 50 miles. For a thousand miles west from Columbus, Ohio, the average rainfall drops about 25 inches, or an inch in 40 miles, but for the growing season the decrease is only about an inch in 100 miles. The fine details of the boundary of a group of species are therefore more dependent on local environmental conditions and on competition than on climate, although in a broad sense it is the climate which is basically responsible.

It is difficult to present in words an adequate explanation of this interplay of general and local environments and their role in determining the limits of range of a set of species. It will be equally or even more difficult for the reader to appreciate it unless he has an opportunity to see it for himself. Naturally that can be done only near such a limit, where two different sets of species are in contact.

No matter where one lives in the United States, he must realize that the native plants about him are adjusted to the climate of his locality and also to those local conditions which are dependent on soil, topography, and competition of the plants with each other. No matter in what direction one may travel, he must realize that there is a gradual variation in the amount and seasonal distribution of heat or rainfall or both, and probably many abrupt variations in soil and topography. He will also see in his travels that some of these local variations tend to neutralize or to compensate for the change in climate, while others tend to add to its effectiveness. As one travels west across Nebraska on the Union Pacific or on U.S. Highway 30, with the annual rainfall steadily decreasing, he may see examples of this from his train window or car. Most of the land is open treeless range, due primarily to the low rainfall. Along the low flat flood-plain of the Platte River, soil moisture is more abundant and trees are growing, while frequent outcrops of rock are too arid for the range grasses and from the road appear wholly barren.

There are some places in the country where the climate changes rather abruptly. These are all associated with mountain ranges, which often have a higher rainfall than the adjacent lowlands and on which temperature decreases with altitude. There is no place where this can be seen better than along the eastern slopes of the Sierra Nevada. These mountains have high precipitation, mostly as winter snow which may linger all summer. The eastern slope is steep and descends into a desert with ten inches of rain or less, and, especially toward the southern part of the range, with excessively high temperatures.

Fig. 14.11. Hepatica. The name comes from a fancied resemblance of the leaf outline to that of the human liver. (Photo by Hugh Spencer, from National Audubon Society)

Fig. 14.12. Indian turnip, or jack-in-the-pulpit. (New York Botanical Garden photo)

Fig. 14.13. Trillium grandiflorum. Several species of *Trillium* are common in the eastern United States. (New York Botanical Garden photo)

Fig. 14.14. Wild ginger (*Asarum canadense*). (Photo by Hal H. Harrison, from National Audubon Society)

From the summit of Mt. Whitney to Death Valley, the descent is almost 15,000 feet, the average temperature rises almost one degree per mile of horizontal distance (about 70 miles), and the rainfall decreases about one inch per mile. Naturally, the change in plant life is correspondingly abrupt. Nevertheless, the local distribution of forest and desert depends on local conditions, the forest trees extending lower and into a drier and hotter climate in favored spots where more soil moisture is available, the desert plants ascending higher and into a cooler and moister climate on rocky slopes with little available soil moisture, excellent drainage, and greater exposure to the sun.

All over the world, wherever plants grow, we find these great groups of many kinds of plants with more or less coincident ranges. All over the world there are exceptions to an exact coincidence, since plants do not act cooperatively but every kind acts for itself. Consequently, some kinds have ranges which project a bit beyond the average boundary, while others fall short. All over the world, the cause of these coincident boundaries, these joint ranges, is a general demand of the plants for a particular kind of climatic environment. All over the world the details of the boundaries of each group of species of plants are set by details of soil and topography and local competition.

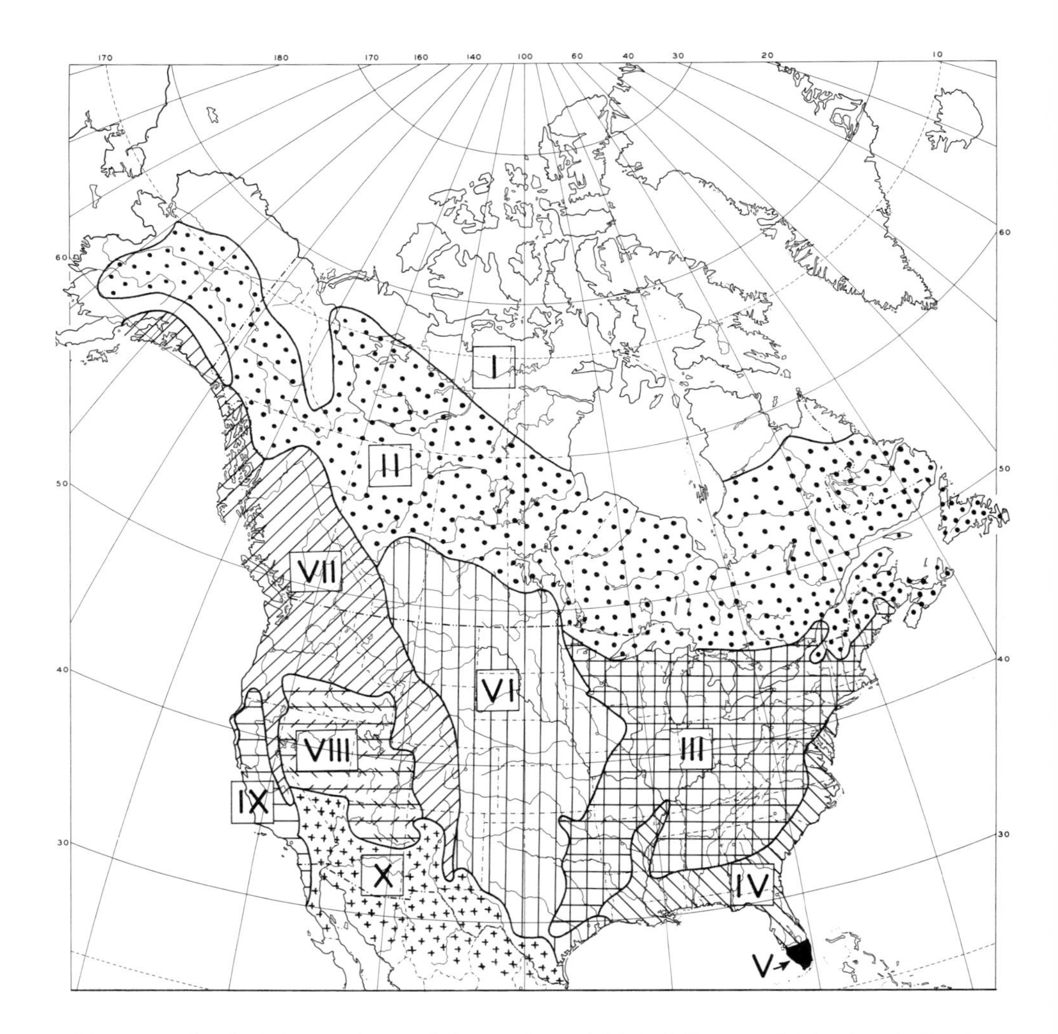

Fig. 15.1. The floristic provinces of the continental United States and Canada. I, Tundra Province; II, Northern Conifer Province; III, Eastern Deciduous Forest Province; IV, Coastal Plain Province; V, West Indian Province; VI, Grassland Province; VII, Cordilleran Forest Province; VIII, Great Basin Province; IX, Californian Province; X, Sonoran Province.

For this map the lines between the provinces have been drawn boldly, so as to show the general outlines rather than the ultimate details. The actual boundaries are in general not sharp; instead they overlap and interfinger extensively, and small enclaves of one province may be wholly surrounded by another.

15/ Flora, Floristic Group, Province

For illustrations of groups of species of similar range, we chose some plants of the eastern United States. We could just as well have chosen other groups from any other part of the world for which reliable information is available. It might have been a similar number from western Europe, and we would have found that they too had, in general, a similar common range, extending north into southern Scandinavia and the Baltic States, east to the Ukraine and Hungary, and south to the Alps and the Pyrenees. They might have been chosen from the Amazon Valley, and we would have found that these plants occupy the lowlands of that valley and much of the Orinoco Valley as well and extend up to an altitude of about two thousand feet in the foothills of the Andes. In short, the plants of the whole world may be classified into groups of associated species, each group occupying its own distinctive territory.

This brings us to the word *flora*, which has been carefully avoided so far. The word is often used loosely in nontechnical writing, but in plant geography it has a very definite meaning. It refers to the kinds of plants, taken collectively, which occupy a specified region. It is a very useful word. Thus we might say, "The floras of western Europe and eastern America are similar." In both regions are various species of maple, oak, beech, elm, and ash, and one familiar with the plants in one continent would feel fairly well at home in a forest of the other. Or we might say, "The floras of Michigan and Australia are very different." In the one are oak, hickory, maple, and ash; in the other acacia, myrtle, and eucalyptus.

We have noted that there are in all parts of the world groups of species which have, in general, the same geographic distribution. Therefore there are also, in all parts of the world, areas of land which have, in general, a uniform and distinctive flora.

The reader may observe that we have here two concepts, the area and the flora which occupies it. Let us call each of these major areas, into which the whole land surface of the world may be divided, a *province*. Let us call the flora of each province a *floristic group*. Every floristic group is a flora, but not every use of the word flora implies a floristic group. We can, if we wish, speak of the flora of a single county or even a single woodlot or bog, but these floras would include only a fraction of a floristic group. We can also speak, quite appropriately, of the flora of the state of Colorado, which may be classified among three or four floristic groups.

In the United States (excluding Hawaii) there are ten such floristic groups and consequently ten provinces. Some of these extend over the boundaries into Canada, Mexico, or the West Indies, and, indeed, only two are wholly confined to our country. We shall presently discuss some of the characteristics of a province and a floristic group, and still later consider each one of the ten in some detail, but for the present it must suffice to list them by name, state their general location, and mention a very few of their most conspicuous or important or better known kinds of plants.

1. The Arctic or Tundra Province: around the world in arctic latitudes; poorly represented in the United States (except Alaska) and developed only above timberline in the mountains of the western states and New England. Moss campion (*Silene acaulis*), mountain heath (*Phyllodoce*), Cassiope.

2. The Northern Conifer Province: from Labrador and Maine westward and northwestward to Minnesota and Alaska. Tamarack, balsam fir, black spruce.

3. The Eastern Deciduous Forest Province: from Maine to Minnesota, and southward across the eastern states to northern Georgia, northern Louisiana, and eastern Texas. Sugar maple, beech, basswood, various oaks and hickories.

4. The Coastal Plain Province: Coastal Plain of the Atlantic and Gulf states and the Mississippi embayment. Bald cypress, live oak, longleaf pine.

5. The West Indian Province: southern Florida and the Florida Keys, and throughout the West Indies. Mangroves, various kinds of palms, gumbo limbo.

6. The Prairies and Plains or Grassland Province: east of the Rocky Mountains from Alberta to Texas and eastward through the present corn belt to Iowa, with outliers in Illinois. Buffalo grass (*Buchloe*), grama grass (*Bouteloua*), Indian grass (*Sorghastrum*), bluestem (*Andropogon*).

7. The Cordilleran Forest Province: mountains of the western states and Canada from Alaska to California and New Mexico. Douglas fir, ponderosa pine, lodgepole pine, western red cedar, western hemlock.

8. The Great Basin Province: most of Utah and Nevada and parts of adjacent states. Sagebrush (*Artemisia tridentata*), greasewood (*Sarcobatus*), shadscale (*Atriplex*), piñon pine.

9. The Californian or Chaparral Province: California, extending into

southwestern Oregon and northern Baja California. California live oaks, digger pine, manzanita (*Arctostaphylos*), madrone.

10. The Sonoran Province: hot deserts and valleys from southern California to Texas and southward into Mexico. Creosote bush (*Larrea*), giant cactus, mesquite, ocotillo (*Fouquieria*).

At least three other provinces and floristic groups exist in Mexico and Central America, making at least thirteen for North America. In South America there are also twelve or thirteen; but Europe, a much smaller continent, has only six. The number in the whole world is apparently about a hundred.

The difficulty in the delimitation of provinces and their associated floristic groups lies not only in the complexity of the facts, which must be about equally obvious to all observers, but also in our personal interpretation of these facts. This is a matter of classification, the fundamental nature of which has already been discussed. To illustrate the matter, let us consider a hypothetical problem of political classification. Let us assume that the state of Texas, which was once an independent nation, had never organized its land into counties. There are no county seats, because government is centralized at the state capital. There is a state board of education, but no county superintendent of schools; there is a state police system, but no county sheriffs. Now the state proposes to divide its land into counties and calls on *you* and *me* to study the situation individually and to make independent reports for the legislature to consider. It is almost inconceivable that you and I by a miraculous coincidence should make identical recommendations, dividing the state into the same number of counties and placing their boundaries in the same position, even though we are working with the same body of facts, the same land, and the same people. It is equally inconceivable that either you or I would recommend the same number (254) of counties as are now in the state.

Now the facts of plant distribution in the United States are pretty well known, but the interpretation of these facts, the assortment of the plant life into floristic groups occupying certain provinces, is a matter of personal opinion. The author recognizes ten in continental United States, but it is only fair to say that other students have arrived at different numbers, making some provinces larger and others smaller. The author admits the truth of the evidence upon which they have reached their different conclusions, but naturally believes that his classification into ten great floristic groups is the most useful in comprehending the plant geography of the nation. There is no need, in an introduction to plant geography, to state these varying ideas, and therefore no need to debate their validity.

Fig. 16.1. Giant cactus, or saguaro, in Arizona. (U.S. Forest Service photo by Rex King)

16/ The Nature of Floristic Groups and Provinces

Let us remember that the geographical history of a species is one of almost continuous migration—sometimes forward into new areas enlarging its range, sometimes retreating and contracting its range. The history of a floristic group is merely the sum of the histories of its component species, and the province is merely the land which it occupies today. There are, however, certain other features of floristic groups which must be presented.

In every floristic group there are a large number of species, usually half to three-fourths of the total flora, which occur in no other province. These plants are so similarly adjusted to all the various environmental conditions existing within the province that they have not been able to migrate beyond its limits or (except possibly at the boundaries) to associate with plants of other floristic groups. These plants are said to be *endemic* in the province. The sugar maple (*Acer saccharum*, Figs. 2.3, 5.4, 14.5) is endemic to the Eastern Forest Province; the giant cactus (*Carnegiea gigantea*, Fig. 16.1) to the Sonoran Province. All the species in our ten provinces have and have had the opportunity to expand their ranges throughout the province. Many of them have done so. Those species which now extend to the borders of the province have also had, and still do have, opportunity to migrate still farther into other provinces. The fact that they have not done so shows how similarly they are all adjusted to one general type of environment. It is a pretty fair assumption that they have all lived together for a long time as members of the same floristic group.

Sometimes, there is evidence or a good presumption that plants actually originated in the province which they now occupy. Such plants are the "native sons" of their floristic groups and may be called *autochthons* or described as *autochthonous*. A full knowledge of the present range of the species shows at once whether it is endemic or not in its province, but evidence of autochthonism is usually only circumstantial.

Much has been written about endemism, and some of the conclusions are interesting, although all of them are just what might be expected from a priori reasoning alone. In general, the larger the area to be considered, the greater is the amount of endemism, and the more effective the barriers which prevent free migration into and out of an area, the greater will be the endemism in the area. The continental islands, that is, those which lie close to a continent and are geologically related to it, have very low endemism. Such islands, for example, are Great Britain and Long Island. This is to be expected, since most of them are within migrating distance of the mainland now and have often been connected with it in the not too remote past. Oceanic islands, on the other hand, which are quite independent of the mainland in distance and history, have a high degree of endemism. Australia, as a continent, has a great number of endemic species, and New Zealand, as a group of larger islands, has an extraordinary number. The widely scattered island groups of Polynesia are of great interest to the phytogeographer. The individual islands in each group consist typically of a central mountain mass surrounded by lowlands. The latter have been built recently by corals and seaweeds. The shores are the haunts of many birds, which commonly travel long distances and are known to be effective agents in the migration of shore plants. Consequently the shores, even in different groups of islands, are populated largely by the same kinds of plants. Ocean currents are also effective in homogenizing the shore floras, if the distances between the islands are not too great. But, even in the strand flora, evolution is constantly at work, and even the commonest and most widely distributed species often show local variation as the result of local evolution. On the other hand, there is no effective method of seed migration between the mountains of the various groups of islands, and there evolution has gone on independently in each group for long periods of time. If the floras of all the Polynesian islands are considered together, the families of plants are found to be the same, and the genera more or less the same, so that phytogeographers commonly regard the area as a single province. But from one island group to another the species vary greatly, and each island has a number of endemic species. Here we can say with some degree of assurance that the endemic plants are also autochthons.

In a continent, North America for example, a considerable fraction of the endemic species of each floristic group are distributed throughout the province. The thirty species already used to illustrate joint ranges were almost all of that type. On the other hand, we remember that climatic conditions always tend to vary more or less from one boundary of a province to another, and

Fig. 16.2. Umbrella tree (*Magnolia macrophylla*) in flower and fruit. (U.S. Forest Service photos by W. D. Brush)

it is understandable that certain species, more delicately adjusted to certain features of the climate or soil, may be restricted to one part of the province. Also, the past climate of the region now constituting a province may have been subject to major change, as in the glaciated parts of North America. This change may have been so drastic that some plants were forced to retreat from a part of the province, and it may have been so recent that they have not yet had time to migrate back.

These matters require some examples for illustration, and for them we turn again to the Eastern Deciduous Forest Province. In it there is a considerable difference in temperature from the long growing season and mild winter of the southern states to the short summer and severe winter, possibly with heavy snow, along the northern boundary. Consequently we are not surprised to find a considerable number of species restricted to the southern half or third of the province, or extending northward to various distances but not to the northern boundary. Such are the tulip tree (*Liriodendron Tulipifera*, Figs. 4.5, 14.6, 16.13), the umbrella tree (*Magnolia macrophylla*, Fig. 16.2), the papaw (*Asimina triloba*), and the silver-bell tree (*Halesia carolina*, Fig. 16.3). Others, such as the swamp white oak (*Quercus bicolor*, Figs. 16.4 and 16.5), apparently prefer the cooler climate of the northern half and do not extend far to the south. We can not be sure that the difference in climate is the sole and direct cause of this partial range unless we know by trial that the plant is not hardy beyond its present range. Competition for space with other species may have a great deal to do with it, but here the climate may be only one step removed as the controlling factor. It happens, also, that the northern half of this particular province was covered by ice during the glacial period. It may well be that the northward migration of some of these southern plants, after the disappearance of the ice, has not been fast enough to spread them over the entire area. In general, we do not definitely know whether the cause of partial distribution in any particular instance is due to present climate or to past history.

Within this same province, the soils along the northern boundary, those in and east of the Appalachian Mountains, and those in the southern states are usually acidic. The soils of the corn belt and the Ohio River Valley, with local exceptions, are mostly calcareous. There are plants which are restricted to calcareous soils and others which are restricted to acidic soils or at least prefer them. Much of the acidic part of the province is characterized by hemlock (*Tsuga canadensis*, Fig. 16.6), chestnut oak (*Quercus Prinus*), and in the near past also by chestnut (*Castanea dentata*, Fig. 8.1), which has recently

Fig. 16.3. Silver bell tree (*Halesia carolina*). (U.S. Forest Service photo by W. D. Brush)

Fig. 16.4. Swamp white oak (*Quercus bicolor*) in Virginia. (U.S. Forest Service photo)

Fig. 16.5. Swamp white oak, leaf and acorn. (U.S. Forest Service photo by W. D. Brush)

been essentially exterminated by chestnut blight. It is also characterized by rhododendron and numerous other shrubs of the heath family (Fig. 16.7). All of these are lacking in the extensive calcareous soils of the Middle West. There we find another group of species scarcely occurring farther to the east, north, or south. Among them are the honey locust (*Gleditsia triacanthos*), Kentucky coffee tree (*Gymnocladus dioica*), buckeye (*Aesculus glabra*, Fig. 16.8), and hackberry (*Celtis occidentalis*); among the herbs are the wake robin (*Trillium recurvatum*), waterleaf (*Hydrophyllum appendiculatum*), and blue-eyed Mary (*Collinsia verna*). Many other species also show this type of partial distribution, but with them grow all those widely distributed species used for mapping in an earlier chapter.

Every other floristic group, no matter where located, presents many similar instances of partial distribution, always against a background of widely distributed species. For example, in the Coastal Plain Province of the southeastern states, the longleaf pine (*Pinus australis*, Fig. 16.9) grows in the Gulf States and extends up the Atlantic Coast as far as Virginia, but does not extend north in the Mississippi Valley. The pecan (*Carya illinoensis*, Fig. 16.10) also lives in the Gulf States, extends up the Mississippi Valley as far as Iowa, but is absent from the Atlantic Coast (except in cultivation). The bald cypress (*Taxodium distichum*, Fig. 9.7), in contrast, grows throughout the province and extends north along the Mississippi Valley to southern Indiana and southern Illinois, and along the Atlantic Coast as far as New Jersey.

Extreme examples of partial distribution also exist in every province. Such, for example, are *Conradina verticillata*, a recently discovered plant of the mint family, restricted to the Cumberland Mountains of Kentucky and Tennessee; *Geocarpon minimum*, a tiny plant known only from southwestern Missouri; and *Thismia americana*, known only from a wet prairie at Chicago. More striking examples are to be found in the west, as the redwood (*Sequoia sempervirens*), limited to a narrow strip along the coast from southern Oregon to central California; the Giant Sequoia (*Sequoiadendron giganteum*, Fig. 1.2), growing only in a few groves in the Sierra Nevada; the Torrey pine (*Pinus Torreyana*, Fig. 16.11), limited to one grove on the California mainland and to one offshore island; and the Gowen cypress (*Cupressus Goveniana*), confined to only a few acres near Monterey, California.

Every one of these plants, notwithstanding its localized distribution, is a genuine member of its floristic group and, in fact, endemic to it. Every one lives closely associated with other species of much wider range. We rarely *know* why they are restricted, although three possible causes have already been

Fig. 16.6. Hemlock. This is the eastern American species, *Tsuga canadensis.* (U.S. Forest Service photos, the detail by W. D. Brush)

Fig. 16.7. Mountain laurel (*Kalmia latifolia*), a common eastern American shrub of the heath family, in North Carolina. (left, U.S. Forest Service photo by Daniel O. Todd; right, New York Botanical Garden photo)

Fig. 16.8. Ohio buckeye (*Aesculus glabra*). (U.S. Forest Service photo by W. D. Brush)

Fig. 16.9. Longleaf pine in Mississippi. This is one of the most valuable species for the production of turpentine, as well as being an important timber tree. (U.S. Forest Service photo by B. W. Muir)

suggested. There is also a possibility that the species originated right there and at a fairly recent time, so that it has not yet had time to achieve a wider distribution. That might account for the little mint, *Conradina verticillata*, and for the Gowen cypress (although it probably does not). It may account for some of the numerous locally restricted species of lupine and penstemon in the Pacific States or some of the many local kinds of hawthorn (*Crataegus*) in the east. It certainly will not explain the distribution of redwood or giant sequoia, both of which are old species, which apparently had much wider ranges in the past. If one knows something of the past history of the floristic group, and especially if one knows the distribution of closely related species, he may produce an explanatory theory which will be satisfactory to himself and may even appeal to a few others.

Not far back, it was explained at some length how groups of species not only live together but also tend to migrate together. That is true. But plants do not migrate to the same place *because* they live together, but solely because they are adjusted to the same environment. Their joint migration may be compared to your trip to a movie theater where you sit beside a total stranger. You do not go there because he does. You both go because you both want to see that movie. This is the first premise to introduce a new topic.

The second premise is: If in a floristic group there are species so narrowly adjusted to environmental conditions that they are able to live in only part of a province, as the redwood in the coastal fog belt of California, then conversely we may expect to find some other plants so tolerant of environmental differences that they can migrate far and wide and become a part of two or even more floristic groups.

Actually there are many such plants, and we need no better examples than two common trees of our country. The quaking aspen (*Populus tremuloides*, Figs. 2.5, 13.2) and the paper birch (*Betula papyrifera*, Fig. 16.12) are widely distributed from the Atlantic to the Pacific in the Northern Conifer Province and the Cordilleran Forest Province. They are equally members of both provinces. Most floristic groups include many plants of this type and they often pose interesting and very frequently unsolvable problems for the phyto-geographer. How shall we account for the tulip tree (*Liriodendron Tulipifera*, Figs. 4.5, 14.6, 16.13) of our eastern states reappearing (as a variety, to be sure) in China? We could settle all such problems if we could only turn back the pages of time and trace all the migrations of such plants back to their origin.

Every floristic group has its geographical limit; every province has its boundaries. These boundaries are quite unlike the accurately defined and surveyed boundaries which mark political provinces or states. Always they

Fig. 16.10. Pecan tree in western Florida. (U.S. Forest Service photo by J. D. Morriss). Right, leaf and fruit. (U.S. Forest Service photo by W. D. Brush)

Fig. 16.11. Torrey pine in California. (U.S. Forest Service photo by H. N. Wheeler)

Fig. 16.12. Paper birch. (U.S. Forest Service photos: Left, in Wisconsin, by P. Freeman Heim; right, in Alaska, by H. J. Lutz)

Fig. 16.13. Tulip tree. (U.S. Forest Service photos: left, twig with unfolding leaves, by E. R. Mosher; right, fruits, by W. D. Brush)

depend broadly on climate, either temperature or rainfall or both, or on geological history; always the details depend on local environment. Since we have good information about climate and geology, we can easily show the general location of a boundary on a map, even on one of small scale. Since local environments may vary greatly within a short distance, the details of a boundary can be shown accurately only on a large-scale map. Sometimes the boundaries are very complicated; there may be islands of one province, each with the characteristic species, completely surrounded by another province, or long tongues of one province projecting into another. Sometimes the actual detailed boundaries are very sharp, and a person may pass from one floristic group into another, that is, from one province into another, in a few steps. In other places the species of the two floristic groups actually mingle in the same habitat along the contact of the provinces.

The sharpest and clearest boundary in the country, so far as known to the author, is the one separating the Eastern Forest from the Prairies and Plains in Illinois and Iowa. A good example of it probably no longer exists in Illinois, but it can still (1963) be seen excellently in the state park just north of Hawarden, Iowa. The forest here is dense and continuous, as far as it goes. Coming out from it to its margin, one penetrates the last tangle of shrubbery (mostly sumac) and in five steps is out on open, treeless prairie and surrounded by an entirely different group of plants. Behind him are various kinds of trees and may apple, yellow violet, pennyroyal, and many other kinds of herbs which he could see also in Maryland or Massachusetts. Before him are various kinds of grasses, and also compass plant (*Silphium laciniatum*, Fig. 16.14), prairie coneflower (*Ratibida columnifera*), prairie clover (*Lespedeza leptostachya*), prairie violet (*Viola pedatifida*), and many other species which represent an entirely different floristic group.

In contrast to this narrow and sharp boundary, consider conditions in northern Minnesota, where the Northern Conifer Province meets the Eastern Deciduous Forest. Martin Grant has described the forest of Itasca County, where the balsam fir of the Northern Conifer floristic group associates in the same habitat with the sugar maple and basswood (*Tilia americana*) of the Eastern Forest. Some seventy miles to the south the balsam fir disappears; about the same distance to the north the maple and basswood disappear. The boundary between the two provinces, as far as these three species are concerned, is therefore about a hundred and forty miles wide. Throughout the corn belt, the boundary between forest and prairie was always sharply defined, as described above. But the two floristic groups divide the land between them over most of Illinois and Iowa and parts of Indiana and Nebraska. Here the boundary

Fig. 16.14. Compass plant (*Silphium laciniatum*). Two views, showing how the leaves tend to be aligned in a single plane. This plane is consistently oriented in a more or less north-south direction, whence the common name.

between the two provinces, where the floristic groups share the land but scarcely mingle, is a good five hundred miles wide from east to west.

We see then that we cannot expect the boundary between two provinces to be always sharply defined or narrow, nor can we expect that the species of the two floristic groups will never mingle. Nevertheless, it is a fact that the floristic groups of the United States are distinct and conspicuously so, as any one can see in his travels. Probably a full realization and appreciation of their distinctness can come only through travel and observation. The reader may be willing to accept the author's word for it, but he will really know it if he sees it for himself.[1]

If a single species can extend its range or be forced to contract it, it is obvious that associated species may extend or contract their ranges at the same time and to about the same extent. The size of the province may therefore change. Its present boundaries may not coincide with those of some past time, and the future may change them again. There is evidence that most of our American provinces have changed their boundaries considerably and that some would be changing rapidly today if agriculture had not intervened.

Lastly, in a general consideration of provinces and floristic groups, it must be stated that there are no universally accepted names for any of them. Most of the names in use have some reference to their location (as the Sonoran Province), to the prevalent type of plant life (as the Grassland Province), or to both (as the Eastern Deciduous Forest Province).

A brief summary of the chapter: The plant life of the whole world may be divided into floristic groups of species which are associated over an area called a province. Each province is characterized by a particular type of climate or more rarely by a particular geological history. Boundaries of a province are commonly broad and are to be described as a belt rather than a line. Along the boundary, details of location are fixed by local conditions, and in some places two different floristic groups are associated. Each floristic group contains many endemic species, which give character to the group and lead us to distinguish one group from another. Some of these endemic plants, because of narrower adjustment to environment (or possible other causes) are restricted to one part of the province. In every province there are some species which, because of broader adjustment to environment, extend beyond its boundaries and are members of two or more floristic groups. Despite these exceptions, the floristic groups of the world are wonderfully distinct.

[1] A student in one of the author's classes listened attentively to the lectures on plant geography. The following summer he motored across the continent, going by one route and returning by another. When he returned to college, he burst into my office and cried, "Dr. Gleason, you know what you told us about the provinces in the United States? *Well, it's all true!*"

Fig. 17.1. Ponderosa pine forest near Darby, Montana. The well scattered trees and lack of undergrowth are typical of many such stands. (U.S. Forest Service photo)

17/ More about Floristic Groups

The question of how a province changes its boundaries and a floristic group extends its range came up in Chapter 13, where the reader was referred to this place.

There are many places in the country where this process of change may be seen, and it is always found to be composed of various small detailed actions going on simultaneously. Since most provinces are characterized by particular types of climate, these detailed actions are usually correlated in some way with climatic change or fluctuation. By fluctuation, we understand the rapid changes of short duration, which we usually refer to as weather instead of climate, but we also include under the term the cyclic changes which affect a short series of years. Thus, there is good reason to believe that the known cycle of sunspot intensity is the cause of certain cyclic fluctuations in climate, lasting eleven or possibly twenty-two years. The ordinary changes in weather are entirely too short to have much effect on plant distribution, but cyclic variations involving several consecutive years may have a visible effect.

Many years ago Francis Ramaley and G. S. Dodds reported on the effect of a series of unusually dry or unusually wet years along the eastern base of the Rocky Mountains in Colorado. Here the Prairie and Plains floristic group comes into contact with the cordilleran conifers, the latter represented at the actual margin chiefly by ponderosa pine (Figs. 9.2, 17.1), which, like nearly all pines, has winged seeds. During a period with rainfall above normal, young pine seedlings spring up for some little distance out on the grass-covered plains and grow as long as the rainfall remains high. With the return of normal or subnormal rainfall, these little trees die and the *status quo ante* is reestablished. Now these changes in rainfall involve at most only a few years, while the life of a ponderosa pine extends at least to two centuries, and in fully established trees the roots go down to a considerable depth. The mature trees therefore reach down to a more permanent water supply and are very resistant to periods of drought, even if the seedlings are not. So, a permanent advance of the forest onto the plains requires a period of above-normal rainfall long enough to enable the young trees to reach a considerable size and great enough to supply them with underground water. The present margin of the forest probably marks the last spot at which these necessary conditions were met within the past two centuries.

But how about an advance of the plains plants into territory now occupied by forest? In a period of subnormal rainfall the large trees are relatively little affected, for they have deep roots to tap ground waters. Only herbaceous plants will be jeopardized, since they depend more on water from the upper layers of soil. In any ordinary period of subnormal rainfall, the forest trees remain unharmed (although possibly retarded in their rate of growth) and the effect of reduced rainfall is shown by the partial replacement of the herbaceous plants of the foothills by those of the plains. With a long or permanent reduction of the rainfall to a point below the limit of tolerance of the pines, the herbaceous plants will die out completely (or almost so), the trees will succumb one by one (thereby reducing the shade and further hampering the forest herbs), and the whole floristic group of the plains will gradually move up into the mountains. We do not know how much the forest margin may have advanced and retreated again during the course of many centuries.

A similar state of affairs may be seen along U.S. Highway 395 as it runs north and south on the east side of the Sierra Nevada in California. Here the contact is between the lower edge of the Cordilleran Forest, again represented by ponderosa pine or by its analogue the Jeffrey pine (*Pinus Jeffreyi*), and the sagebrush (*Artemisia tridentata*, Fig. 17.2) of the Great Basin. Here can be seen areas of open forest with widely spaced trees of pine, all of considerable size and bearing cones, while the undergrowth is mostly a tangle of the bushy gray sagebrush. One immediately gets the impression that these trees are the last rear guard in the retreat of the forest farther back into the mountains, while the sagebrush, already in control of the ground, is just waiting for these last few trees to die and leave the desert plants in undisputed possession.

Along the northernmost border of the forest in Alaska, thousands of young spruce trees are appearing in the treeless arctic wastes beyond the tree line. Here the controlling factor is temperature, and the presence of these young trees suggests that the temperature is growing milder. No one knows whether this is a cyclic change only, soon to be followed by a return to normal temperatures and leaving thousands of dead trees out on the tundra, or whether it is part of a permanent climatic change which is resulting in the gradual extension

Fig. 17.2. Typical sagebrush (*Artemisia tridentata*) plant near Pocatello, Idaho. This characteristic shrub of the Great Basin Province holds its small, gray-green leaves all winter. (U.S. Forest Service photo)

Fig. 17.3. Joe-pye weed (*Eupatorium maculatum*), a characteristic species of moist, low, sunny places in the Eastern Deciduous Forest Province. (Photo by D. M. Compton, from National Audubon Society)

of the forest toward the north. We do know that there has been much shrinkage in the Alaskan glaciers in the last half century and suspect that this forest advance is the effect of the same general increase in temperature.

The Northern Conifer Province is characterized by great numbers of peat bogs, each with its mat of sphagnum moss or bog sedge, and with its typical zone of shrubs, such as leatherleaf, Labrador tea, and bog rosemary. It is generally known that this whole province is correlated with cold winters and cool summers, and that it is yielding to the advance of the Eastern Deciduous Forest from the south. Undrained acid bogs of this type are now found in the latter province, but the bogs are being replaced by the wet-land forest of the eastern states. The southernmost bogs now stand surrounded by trees of the Eastern Forest, which has long since driven out the pines, balsams, and spruces of the northern forests, but these little bogs still persist, islands of northern species in the midst of a more southern forest. Let's see what is happening to them in a sort of mopping-up process leading to the complete occupation of the land by the eastern forests.

We have already discussed how ponds and lakes tend to fill themselves by deposition of silt and accumulation of muck, and how the zones of plants move in toward the center and one by one disappear. One by one, a number of eastern species migrate into these bogs. Probably they could not migrate successfully into a similar bog in Labrador, where the climate is still much too cold, but they can in Illinois or Ohio or New York. The white water lily (*Nymphaea odorata*) appears in the open water, the water willow (*Decodon verticillatus*) at the edge of the pond, the blueberry (*Vaccinium*) and the swamp dogwood (*Cornus Amomum*) among the shrubs; colonies of joe-pye weed (*Eupatorium maculatum*, Fig. 17.3), cardinal flower (*Lobelia Cardinalis*), boneset (*Eupatorium perfoliatum*), and monkey flower (*Mimulus ringens*) develop wherever they can find space. All of these and other eastern species appear by the accidents of migration. When the forest finally closes in on the bog, it is not the northern forest of black spruce and tamarack, but an eastern forest of silver maple, red maple, elm, and ash.

The greatest struggle between floristic groups in the United States has been waged between the eastern forest and the flora of the prairies and plains. It has been fought back and forth from Ohio to Nebraska and has continued for some thousands of years, from the retreat of the glacial ice to the arrival of the white man, who eventually stopped or greatly modified the fight. In general the Prairies and Plains Province is characterized by lower rainfall, especially during the winter months, than the Eastern Forest Province, and by hotter summers and colder winters.

The early stages of the battle were never seen by a botanist, because they took place too long ago. Several thousand years ago the climate of the Prairies and Plains Province was apparently considerably drier than at present and also probably somewhat warmer. This has led to our modern name for that time, the *xerothermic period*, from the Greek roots *xero*, meaning dry, and *therm*, meaning heat. Such conditions gave an advantage to the prairie floristic group, and it migrated eastward, going certainly as far as Ohio and probably reaching even into New York.

After the xerothermic period the rainfall increased; the climate became more favorable to the forest; and the forest began an advance toward the west, almost completely exterminating the prairies from Ohio and Michigan. The forest came to occupy probably ninety percent of Indiana, probably fifty percent of Illinois, and was making rapid progress elsewhere, when an unexpected event happened. The Indian arrived and started the custom of setting fire to the prairie whenever the grass was dry enough to burn, which was usually in the pleasant months of autumn. The reason for the practice seems to have been that the fires drove deer and other game animals into the forest where they were more easily stalked and killed by the Indians for food.

We do not know when this period of repeated fires began; it may have been about 5,000 years ago. It continued until the white man had begun the settlement of the prairie states, and many observant men a century ago saw and wrote about the effects of the fires on the forest. The fires were commonly driven from west to east by the prevailing west winds. The prairie grasses and herbs were not killed by the fire, since their buds for the next year's growth are underground. Only the dead stems and leaves of the season just closed were consumed. The fires reached the western edge of the forest and burned into it for some distance, feeding mostly on the fallen leaves. They were reduced to slowly burning ground fires, doing little damage to the thick-barked trees and little also to the herbs, whose winter buds are underground. But shrubs and seedling trees bear their buds for next year's growth on the twigs above ground, and they were easily killed. Since the young trees were killed, the forest could not reproduce. With each death of a mature tree, the forest was thinned and more sun-loving prairie plants could grow beneath it, while the shade-loving forest plants found less and less opportunity. Eventually the forest was reduced to mere groves of scattered trees rising above a prairie. These trees also disappeared in due time, and the replacement of the forest was complete.

This process was continued for perhaps 5,000 years, gradually pushing the

Fig. 17.4. Shingle oak (*Quercus imbricaria*) in Maryland. This species, which occurs over much of the deciduous forest region of eastern United States, is also one of the pioneers in the invasion of prairie by forest. (U.S. Forest Service photo by W. D. Brush)

forest farther and farther to the east, but not destroying all of it. Some forest persisted along streams or on the eastern side of ponds and lakes, where the fires seldom reached it. Prairies began to reappear in Ohio; extensive areas of prairie developed in northern and northwestern Indiana. About half the forests of Illinois were destroyed and the prairies correspondingly enlarged until they occupied about 70 percent of the land. In Iowa they came to occupy a still larger proportion, but in that state, being farther west, the forest had not previously made as much progress as it had farther east.

This period of prairie advance was halted by the arrival of the white man in the nineteenth century. The Indians were driven out; the land was gradually put under cultivation; and most of the prairie fires were stopped. They did occur, of course, as long as any remnants of the prairie persisted, but they were few and far between. Now, with a climate suited to forest growth and no fire to prevent this growth, the forests again began a march westward, and this march has also been observed and described by competent persons. The xerothermic condition of several thousand years before had disappeared, and the chief resistance to the forest advance was a tough and dense prairie sod in which seedling trees started with difficulty, coupled with full exposure to the hot sun, dry air, and frequent droughts of midsummer.

The line of attack was two-fold. From the forests which had persisted on the lee (eastern) side of the rivers, the trees crept up the small streams which run from the prairies into the rivers. There the seedlings had a better chance of finding the necessary water. The first trees were those with winged seeds, and the procession was led by cottonwood (*Populus deltoides*), green ash (*Fraxinus pennsylvanica* var. *subintegerrima*), elm (*Ulmus americana*, Fig. 2.4), and silver maple (*Acer saccharinum*, Fig. 13.11). Close behind them came plants with edible fruits, such as wild cherry (*Prunus*) and hackberry (*Celtis occidentalis*), while the slow-moving oaks, hickories, Kentucky coffee tree (*Gymnocladus*), and buckeye (*Aesculus*) brought up the rear.

The other attack was by moving out upon the uplands perpendicular to the streams. There, in the relatively dry soil, the marginal forest trees were mostly hickories and oaks, especially the shingle oak (*Quercus imbricaria*, Fig. 17.4). At the very edge of the forest was a tangle of shrubs, especially sumac (*Rhus*, Fig. 17.5), hazel (*Corylus*), and wild plum (*Prunus americana*). The shade from this strip of shrubs and the overhanging trees weakened the prairie sod. The shrubs marched steadily outward, dying off at the rear where they were shaded by the growing trees. New tree seedlings constantly came up in advance of the parent trees. Herbaceous plants which demand shade were

Fig. 17.5. Common sumac (*Rhus glabra*). One of the pioneers in the invasion of prairie by forest, this shrub actually occurs over much of the United States. (New York Botanical Garden photo)

Fig. 17.6. Grove of white pines in New Hampshire. Although white pine is now most at home in the New England region, isolated stands occur as far south as North Carolina, Tennessee, and Illinois. (U.S. Forest Service photo by John McGinnis)

close behind, and other herbs which need humus were the last to appear. Observant botanists have stated that the advance of the zone of shrubs was about thirteen feet per year. The longitudinal advance along the streams was much more rapid, although no measurements are known, but there are places where trees are growing today more than a mile beyond their extreme limit of only fifty years ago. That would be at the rate of a hundred feet a year, and there is reason to believe that the forest might have moved farther in the same time, and therefore faster, if it had not been hampered by the use of most of the land for agriculture.

It is not to be concluded that floristic groups everywhere move with such speed. In the northeastern states only a few thousand years have elapsed since the retreat of the glaciers, and the change of climate has been profound, permitting and encouraging a correspondingly rapid movement of floras. Movements have doubtless been equally rapid in the recently glaciated areas of Europe and Asia, but in other parts of our country and in most other parts of the world there is little or no evidence of rapid migration of the various floristic groups. They have reached a state of relative equilibrium correlated with the present climates, and there they will stay, except for minor cyclic changes, until geological events alter the climate and start them into new migrations. There is one place, however, in which climatic change is believed to be either in progress or quite recent, and that is along the southern margin of the Sahara Desert in Africa. In this area, there is good reason to believe that forests are being converted into open park-like savanna land and the latter again into desert. In other words, the Sahara appears to be growing toward the south.

Whenever a floristic group is forced to retreat with some degree of rapidity, colonies are often cut off from the main body and persist for a time as isolated outposts, often termed relic colonies. As time goes on, these will probably all be replaced by the surrounding advancing group.

In the eastern states four types of relic colonies are well known. First, the rapid advance of the forests at the close of the xerothermic period cut off many relic colonies of prairie in southern Ohio and Kentucky and some in southern Michigan. These are now mostly occupied by agriculture, but plants of prairie species, still persisting along roadsides, railways, and fence rows, often remain to tell the story.

Second, at the beginning of the period of prairie fires, numerous groves of trees were isolated from the retreating forest, and many still persist. Almost without exception they occupy habitats where they were protected on the west side by water or wet ground; as a result, they escaped the destructive fires. Other groves remained for some time but were finally replaced by prairie before the scientific history of the region began, but so recently that their location is marked today by tracts of typical forest soil. We do not know how many years are necessary to convert forest soils into the black prairie soil. There were doubtless other groves which disappeared so long ago that their soil has been completely changed into prairie soil and no trace remains of their location.

Third, at the close of the glacial period, the eastern forests began a rapid migration to the north. In this movement, colonies of the Northern Conifer group were isolated from the main body, and many of them have persisted to this day. Some of these colonies are groves of white pine (*Pinus Strobus*, Fig. 17.6), such as are found in North Carolina, Tennessee, Ohio, Indiana, and Illinois, usually on outcrops of sandstone rock. The most numerous are the little peat bogs, which have been discussed already, of which there are thousands in Michigan, Wisconsin, and the northern parts of Illinois, Indiana, and Ohio.

Fourth, the forests of the Northern Conifer Province also advanced to the north at the close of the glacial period and in their march cut off fragments of the Arctic Province. These persist today on the tops of the higher mountains of New England and New York, but they are much better developed on the mountains of eastern Canada. In former times there were hundreds of these colonies at lower elevations on bare rocky hills where the soil was so thin that forests had difficulty in getting started. Even today many of them remain unforested. The herbs and shrubs on them are mostly members of the Northern Conifer floristic group, but growing with them are occasional plants of Arctic species, such as some of the saxifrages.

One of these four types of relic colony, the peat bog, was chosen before to illustrate how species of the surrounding province gradually enter it, mix with the relic species, and eventually take over completely. The other three types illustrate the same general features, the incursion and colonization of more and more species from the surrounding province, until in due time the relic species are completely replaced. Relic colonies also illustrate another general principle in the migration of floristic groups: the relic colonies linger behind in the habitats least suited to the invading flora. As a corollary to this, it is also evident that an advancing flora moves forward most rapidly in the habitats best adapted to itself. Human migrations follow exactly the same principles. To the victor belong the spoils; the loser must be content with what is left.

Fig. 17.7. Live oak (*Quercus virginiana*) in a city park in New Orleans. The name live oak is applied to several evergreen oaks, which keep their leaves all winter. (U.S. Forest Service photo by R. R. Winters)

From this chapter and the preceding one we can learn an important feature of floristic groups. These plants not only live together today, but they have also been living together and migrating together in the past. They have a history. They may be compared to a nation among mankind, whose territory expands or contracts as it is the victor or the vanquished in war, or which may even migrate to new territory, as the Goths migrated from their northern homeland into southern Europe. The discovery of this history is one of the most interesting problems of phytogeography, but also the most baffling. There are few lines of evidence for which we may look or from which we may draw plausible conclusions. The location of relic colonies, such as were described previously, gives important clues to the past range of a floristic group, just as the existence of advancing tongues or strips predicts something of its future or potential range.

The most convincing evidence is that obtained from fossil plants. These are most valuable in showing the nature of the flora in remote ages, for which fossils are the only evidence available. Leaves and stems are fossilized after burial in mud or peat and, since they are not often carried far from the parent plants, they tend to give a rather fragmentary record of the flora as a whole.

Fortunately there is one kind of fossil the importance of which was not even suspected half a century ago, and that is the fossil pollen grains which are preserved by the billion in every peat bog. They may be separated from the peat, examined under the microscope, and identified accurately. Peat is deposited in the bogs in successive superposed layers, so that samples of pollen grains taken from all levels from the bottom of the bog to the surface give a picture of the plant life surrounding the bog from its inception to the present time. The pollen is almost all from wind-pollinated species of trees and shrubs, such as oak, hickory, poplar, birch, ash, and all of our conifers. All of these produce huge quantities of pollen, which is blown so easily through the air that a bog can collect pollen from trees at a considerable distance just as well as from the spruce and tamarack growing right in it. Wind-pollinated herbs produce much smaller quantities of pollen, and insect-pollinated plants are, in proportion to their number, very poorly displayed in the fossil record. The study of fossil pollen in the northern states, Canada, and northern Europe has given a very clear picture of the migration of floristic groups since the close of the glacial period. Bogs are developed only in regions of considerable rainfall and humidity. All of our northeastern bogs show, in general terms, three types of flora: first and oldest, a flora of conifers, including fir and spruce, indicating the colder climate immediately following the glacial period; second,

Fig. 17.8. Dwarf palmetto growing under loblolly pine on Cape Hatteras, North Carolina. This is near the northern limit for this species (the palm). (National Park Service photo)

a forest with predominant oak and pine, showing a drier climate during the xerothermic period; and last, a flora more like that of today, indicating a return to a more humid climate.

If a floristic group has a history, it must also have had an origin. No one dares to say that the deserts of Arizona have existed forever, or that Canada has always been covered with forests of spruce and fir. When we consider the origin of a floristic group, there are very few statements that we can make with assurance. One of them is that the process required a long time and that some floristic groups are older than others. If we admit the truth of the latter statement, then we must admit the possibility that some floristic group is in the process of origin right now. Since the effective lifetime of a phytogeographer is at most fifty years, not one of us can say whether such a process is going on now, or where, or why. Since the study of plant geography is less than two centuries old and botany as a science only four centuries, the collected wisdom of botanists can not show that such a development is going on. Nor would two thousand years be sufficient to show it, although it would show significant changes in the size and location of many provinces.

Another statement which we can make is based on the fact that the flora of each province is different from those of adjacent areas. The origin of a floristic group must therefore depend at least partly on the evolution of new species. We can imagine that some of these new species have sprung from parents within the same area and that the parents have sooner or later disappeared, either through too strenuous competition or through environmental changes. Other new species may have arisen in an adjacent province and migrated in, either with or without specific evolution as they came. If we compare the flora of any province with those of adjacent areas, we usually find a number of similarities, not in identical species, although these may exist, but in closely related species or genera. We may therefore consider that a principal source of new species in the development of a new floristic group is differentiation (evolution) from the species existing in surrounding areas, and always in correlation with the change in environment which is, after all, the fundamental cause of the origin of the new floristic group.

The Coastal Plain Province is undoubtedly the youngest in the United States, and, as such, should give us the best chance to speculate on the origin of its flora. The present lands of the coastal plain along the Atlantic Ocean and the Gulf of Mexico were only recently, geologically speaking, still under water. A good share of the coastal plain is even today submersed. In spite of the youth of the land, and hence also of the floristic group which occupies it,

Fig. 17.9. Clump of little bluestem in Somerville County, Texas. This prairie species also occurs in openings throughout much of the eastern deciduous forest region. (U.S. Soil Conservation Service photo by Ben Osborn)

the flora is composed of a large number of species, and all of them must have arrived or arisen there in a comparatively short time, either with or without evolution accompanying their migration. If we examine the flora and seek for other provinces in which the same species, or related species or genera, exist today, we are led to consider four possible sources for its origin.

One striking feature of the coastal plain is the number of species of pine and oak which grow there, and the extent of the forests which they produce. Both of these genera of trees are distinctively of the temperate zone. To be sure, pine extends south into the tropical zone in Central America, Cuba, and the Philippine Islands, but with very few species there, and these most closely related to species of the temperate zone. Oak does not reach the West Indies, but is abundant in the cool highlands of Mexico, while one species reaches Colombia in South America and many exist in the Malayan region. We look naturally to the Eastern Forest Province, just north or west of the coastal plain as the place of origin of the pines and oaks of the coastal plain. Some of the species are unchanged, as blackjack oak (*Quercus marilandica*) or Spanish oak (*Quercus falcata*), or very closely related to northern species, as Ashe's oak (*Quercus margaretta*), while others have no close relatives at all in the eastern forests, as the live oak (*Quercus virginiana*, Fig. 17.7). The sweet gum (*Liquidambar*) and the sassafras are identical in both provinces. The southern red maple (*Acer rubrum* var. *Drummondii*) and the southern beech (*Fagus grandifolia* variety *caroliniana*) are so like the northern trees that they are called merely varieties, while the southern sugar maple (*Acer barbatum*), although much like the northern tree, is usually regarded as sufficiently different to be called a separate species. Many other examples could be cited, but these may be enough to indicate that the principal origin of the coastal plain flora was from the north or west and either with or without differential evolution.

There are also plants of undoubted tropical affinity in the coastal plain. Most conspicuous among them are the palmettos (*Sabal Palmetto*, Fig. 22.42; and *Sabal minor*, Fig. 17.8) which extend northward along the Atlantic coast to North Carolina. As everyone knows, the palm family as a whole is tropical, and these palmettos are among the very few kinds of palm which enter the temperate zone. There are also numerous species of sedges, especially in the genera *Cyperus*, *Scleria*, and *Rhynchospora*, which have no close relatives whatever at the north but are especially well developed in the American tropics. These plants are largely inhabitants of swamps and bogs, and they may well have immigrated, often evolving as they came, across the extensive swamps of southern Florida from the West Indies. Two large genera of

Fig. 17.10. Buffalo grass, one of the most characteristic plants of the short-grass prairies east of the Rocky Mountains. (U.S. Soil Conservation Service photo by Hermann Postlethwaite)

Fig. 17.11. Blue grama (*Bouteloua gracilis*) near San Antonio, Texas. This is one of the commoner species in the more favorable sites on the short-grass prairie. (U.S. Soil Conservation Service photo by Hermann Postlethwaite)

grasses, *Panicum* and *Paspalum*, are especially well represented on the coastal plain, and probably came from the same source.

Some other plants arrived from the tropics by way of the lowlands of Mexico and entered our coastal plain in Texas. The best documented history of plants of this sort is of the ironweeds (*Vernonia*). Not one Mexican species lives also in Texas today, yet the nearest relatives of the Texas species are all Mexican. From Texas they migrated first east and then north, evolving as they traveled, until they occupied all of the coastal plain. From the coastal plain some migrated still farther north, accompanied by specific evolution, into the Eastern Deciduous Forest and Prairies and Plains provinces, while one has crossed the strait and appeared in the Bahamas. Not one has reached the Pacific slope of the United States.

The fourth and least important source of coastal plain species is the Prairies and Plains Province, which seems to have furnished a small number of plants. It is suspected that these are mostly recent immigrants, probably entering the coastal plain during the postglacial xerothermic period.

Notwithstanding the youth of the coastal plain, there are some kinds of plants which may have immigrated at some remote period. Notable among these are the pitcher plants (*Sarracenia*). They are all restricted to the coastal plain except the common pitcher plant (*Sarracenia purpurea*) which extends into the northern states, as has already been discussed. The pitcher plants appear to be of tropical origin, and their nearest relative, the genus *Heliamphora*, is limited to the mountains of British Guiana and southern Venezuela. The tiny curly grass (*Schizaea pusilla*), actually a fern, which lives on the coastal plain in New Jersey, Nova Scotia, and Newfoundland, is a member of an almost exclusively tropical group of ferns. When and by what route its ancestors reached America can only be surmised. And what can we say of the geographical origin and history of the Venus's-flytrap (*Dionaea*, Fig. 22.44), our most remarkable insectivore, now restricted to a small area of wet pine barrens near Wilmington, North Carolina? Nothing.

Second in point of youth is the Prairies and Plains Province, and in it again the location of the nearest relatives of the flora shows that it has been derived chiefly from three sources, either with or without accompanying evolution. Probably the most important of these, and certainly the most important in the eastern part of the province from Nebraska and Kansas east, has been the Eastern Forest Province. Four of the commonest grasses in the prairies of Iowa and Illinois have come directly from the Eastern Forests, where they are still widely distributed. These are the big bluestem (*Andropogon Gerardi*, Fig.

22.51), the little bluestem (*Andropogon scoparius*, Fig. 17.9), the Indian grass (*Sorghastrum nutans*), and the slough grass (*Spartina pectinata*). From the Sonoran Province to the south have come several other grasses, notably the buffalo grass (*Buchloe dactyloides*, Fig. 17.10) and the grama grasses (*Bouteloua*, Fig. 17.11). The Great Basin Province has furnished numerous species of the composite family and the bean family, including various kinds of sagebrush (*Artemisia*) in the former and of milk vetch (*Astragalus*) in the latter.

For these two relatively young provinces, we can pick out numerous plants which we can refer with some assurance to one origin or another. For the older provinces, we can not. We really know very little about their origin. We can think about it, and wonder about it, and theorize about it if we are brave enough, and publish and champion our theories if we are foolhardy enough. Your author has been bold enough to develop some ideas of his own, but he is not foolhardy enough to give them to his readers as an ex-cathedra dictum, telling them just what things have happened over the past few million years. He merely wishes to encourage his readers to appreciate the magnitude of the time involved, to understand the general nature of the processes which must have been involved, and to wonder at the unknown events of the past long eons which have led to the present green cover of the world.

But let us turn from the past, forever hidden behind the clouds of time, and contemplate again the sunlit present, where we have yet to consider still another feature of plant life which has a great deal to do with plant geography.

18/ Vegetative Form

It has already been stated (Chapter 12) that there are three different bases upon which plants can be classified and which will give results of interest or value to the phytogeographer. We have already considered one of these and have discussed the distribution of plants which *live together*. This leads to the recognition not only of small communities but also of large and extensive floristic groups. We now have to discuss the distribution of plants which *look alike*. We naturally expect that different individuals of the same kind of plant will look alike; in fact, we recognize kinds of plants by the similarity of individuals. More important to us now is the resemblance between different kinds of plants.

Every degree of similarity may be observed in nature. Sugar maple and Norway maple look so much alike that many persons do not know them apart. When the leaves are off in winter a red maple looks much like a beech. At a little distance a spruce looks much like a fir. Such statements can be multiplied indefinitely, but it is very doubtful that we could draw any geographical conclusions from them.[1] To get some value from it we must consider broader and more generalized types of similarity; we must arrange plants into a fairly small number of large and comprehensive groups. Otherwise we shall soon be lost in a maze of details.

[1] Readers may wish to make some observations of their own on this subject. If they live in the northeastern states, they may find in upland woods two common plants growing side by side and looking very much alike until they come into bloom; these are the white snakeroot (*Eupatorium rugosum*, Fig. 22.22) and the horse balm (*Collinsonia canadensis*). In moister situations the Virginia knotweed (*Polygonum virginianum*) and the lopseed (*Phryma leptostachya*) are very common, often grow side by side, and look even more alike, even to the flower clusters. In quiet waters of ponds a trio may be found, the water shield (*Brasenia schreberi*), the water smartweed (*Polygonum natans* forma *natans*), and the pondweed (*Potomogeton natans*). In each of these three sets the plants are totally unrelated, beyond being angiosperms. Observers will soon note that such very similar plants almost always grow together, and they may arrive at the conclusion that the chief reason for their similarity is that they live in the same environment.

Fig. 18.1. A forest in Indiana. The large trees are all white oaks. (U.S. Forest Service photo by Lee Prater)

Probably the best way to penetrate this subject is by examples, and we shall use three from different parts of the United States. In an eastern woodlot (Fig. 18.1), as in Virginia or Ohio or Indiana, several different forms of plants may be seen, but the great bulk of them can be placed in three groups. First, there are the trees, plants of great size, living to a great age, and set so closely that their crowns often form a continuous canopy. Second, under the trees are many kinds of shrubs, woody plants which seldom grow more than a few feet tall and rarely form a continuous thicket. Third, there are many kinds of herbaceous plants, some only a few inches tall, others of larger size or even surpassing the shrubs. These three constitute almost the entire plant life of the forest. What is there left which we have not included? On the ground and on the trunks of the trees, mosses and lichens are growing; we can scarcely include them among the ordinary herbs. Here and there, climbing over the shrubs or up the trees, are some vines, as the wild grape or the poison ivy. It might be well to keep them in a group by themselves. On the ground, on fallen logs, and even on some live trees, are mushrooms and shelf fungi. Under the ground are countless billions of soil bacteria and soil fungi, and we may also find some parasitic fungi on the leaves and stems of living plants. All the vines, mosses, lichens, fungi, and bacteria together amount to a very small fraction of the total *bulk* of the plant life, however, although they may be represented by numerous kinds and huge numbers of individuals. We are still justified in saying that an eastern forest is composed primarily of trees, shrubs, and herbs. These represent three very broad, generalized, inclusive types of plant life.

For our next example we go to the cattle ranges of eastern Colorado (Fig. 18.2) and compare it with our eastern woodlot. Of trees there are none. Of shrubs there are almost none, although we may find a few growing around a rock ledge here or there. Vines are lacking. Mosses and lichens are very few. Broad-leaved herbaceous plants are present in a great number of different kinds but with comparatively few individuals of each. The great bulk of the plant life consists of grasses, millions of them, forming an almost continuous sod and stretching away in every direction to the horizon. We may regard the grasses as a fourth vegetative form.

Lastly we go farther west to a desert (Fig. 18.3) in southern Arizona or New Mexico. Here we find no trees, but plenty of shrubs of various shapes and sizes; most of them, in contrast to the shrubs of the eastern woodlot, are low and bushy and have much smaller leaves. Growing among them we find an entirely different kind of plant life; plants like the cacti, with minute or no leaves and thick juicy stems, or like the century plant (*Agave*), with thick juicy leaves. We saw nothing like them in Colorado or Ohio, although we might have found a few had we looked carefully in the right spots. Suppose we have visited the desert just after the warm spring rains. We then find vast numbers of little herbaceous plants which spring up rapidly, bloom after a short period of growth, ripen their seeds, and die. These are not like the herbs of the eastern states. The latter start to grow with the arrival of warm weather and most of them continue to grow until they are killed to the ground by the frosts of autumn. These Arizona plants live only a single season; they start to grow with the rains and die with the drought. They survive until next year only in the seed stage. We can regard them as still another type of plant life.

These examples have been used, not to describe or even hint at the plant life of the three regions, but to enable the reader to appreciate what is meant by broad and generalized groups of plants based on a similarity of appearance and structure and behavior. Several such groups have been mentioned and a few others barely suggested. These are not all that we might distinguish, but they, or at least most of them, are familiar to every observant American. Another type, equally well-known, includes water plants floating or submersed in quiet ponds and rivers. Still another, familiar to all visitors to Florida, includes the so-called air plants which perch upon the branches of trees.

Each of these groups is characterized by a general similarity in the appearance and structure of the component plants. The form, size, and structure of the plants, by which we know them immediately on sight, are always correlated, and usually very closely, with their habits. *Habit* refers to what a plant does and how it behaves, not only at the instant we look at it, but continually. It is not always immediately perceptible. To understand habit requires observation through a whole season or even through the whole life of the plant. A century plant (*Agave americana*, Fig. 5.2) blooms but once at the end of its long life. One might watch such a plant for many years without even suspecting this peculiar and interesting habit. Since the habit and appearance of plants are correlated, the various types of plant life are characterized as much by one as by the other, although they are usually referred to as vegetative forms or life forms.

Botanists have tried to classify the plants of the world into a series of vegetative forms, using both structure and habit as a basis. This has been done in different ways. Any number of forms may be recognized, depending on the breadth of view. For example, we have already spoken of trees and shrubs

Fig. 18.2. Short-grass prairie on open range in northeastern Colorado. The dominant plants are buffalo grass and blue grama. (Photo by H. L. Shantz)

as two distinct types. Maybe they should be combined. After all, about the only difference in structure and appearance is in their size. In habits, both may be found in many different environments, both live many years (some shrubs are short-lived), both include plants with evergreen and deciduous leaves, and, what may be more important, both types add new growth each year to the old growth from previous years. This is quite unlike the habit of the herbaceous plants of our latitudes, which die down to the ground at the end of every season and start next year from seeds or from the basal or subterranean parts of the plant.

This habit of the herbaceous plants of the temperate zone, so well known to all of us, is obviously correlated with the environment. Winter is a violent season in these latitudes and plants must have some protection, not so much from the cold itself as from breakage by wind and sleet, and, above all, from desiccation of the stems and leaves at a season when the lost water cannot be replaced by absorption from the frozen soil. But, you may say, we have evergreen herbaceous plants as well as evergreen trees, even in cold climates. That is correct, but they constitute a small minority in contrast to the great majority of herbs which die down to the ground every autumn. In the tropics, conditions are entirely different. There is often no climatic reason why an herbaceous plant should stop its growth at all. Some of them do not, and a plant which is an herb this year may become a small shrub next year or even a big shrub after several years. Since in our trees and shrubs the growth of next year is added to the growth of this year, the existing stems and branches must be woody enough to provide strength and resistance to wind and sleet and must be covered with a nearly waterproof bark to prevent excessive loss of water.

In other and shorter words, our trees and most shrubs pass the adverse winter season with their buds for next year's growth already developed on the branches and exposed to cold, sleet, and snow, while our herbaceous perennials have their buds below or at the surface of the soil, where they are usually protected by a mulch of dead leaves.

It was probably the observation of this difference in habits which led Christen Raunkiaer into his study of vegetative form. He lived in Denmark, where the winters are cold and the summers reasonably warm. Probably as a result of his location, he based his classification of life forms primarily on the position of the resting buds by which plants pass the adverse season of the year. Unfortunately, he did not take into full account the fact that in some parts of the world growth of plants is continuously possible, or that the adverse season may be due to either cold or drought or both. As a result, he placed in the same vegetative form the trees of the north temperate zone, which live through a winter three to six months long, and those of the Amazon Valley where all seasons are equally warm; into another class he combined the dwarf shrubs so characteristic of polar regions and those equally abundant (not the same species, of course) in hot deserts. If you or I should contemplate the whole plant kingdom and arrange its thousands of species into a similar number of groups, it is probable that our system would differ here or there from Raunkiaer's, and it is possible that we might even improve somewhat on his system.

The value of Raunkiaer's classification of life forms lies not so much in its innate merit as in the use he made of it. That led to important results, to which we shall soon come, and, strange though it may seem, any of several other possible classifications would have led to the same or nearly the same conclusions. First, we must outline his system briefly. He recognized five principal types of life form, and four of these he divided into three to fifteen minor types each, making thirty in all. In the tabulation below not all thirty are mentioned.

A. Phanerophytes (literally, *exposed* plants) are perennial plants one to many feet tall, with their buds (by which they pass the adverse season, if any) fully exposed to the weather. Trees and most shrubs belong here. Among his subordinate groups he had (1) stem succulents, such as cacti, (2) epiphytes, (3) evergreen trees and shrubs, and (4) deciduous trees and shrubs. The last two groups were further divided according to size.

B. Chamaephytes (*ground* plants) are dwarf shrubs, rarely more than a foot tall, with their buds located so near the ground that they are protected by dead leaves or snow. The commonly cultivated periwinkle, also known as myrtle (*Vinca minor*), is a good example.

C. Hemicryptophytes (*half-hidden* plants) are herbs with the buds for next season's growth located at the surface of the ground. Many of our perennial herbs are of this type, and the dandelion is as good an example as any.

D. Cryptophytes (*hidden* plants) are perennial herbs with their buds well below the surface. Raunkiaer had three subgroups: (1) geophytes (*earth* plants) with bulbs, tubers, or rhizomes well below the surface of the soil, (2) helophytes (*swamp* plants) with their buds in the earth at the bottom of shallow water, as cattail (*Typha*), and (3) hydrophytes (*water* plants) with the whole plant body at or below the surface of water, as the water lily.

E. Therophytes (*summer* plants) are annuals, passing the adverse season as seeds.

It is, of course, obvious that the basis of Raunkiaer's classification is not entirely form and appearance, but largely habit and behavior, especially as to the position of the resting buds.

In using his system to reach the conclusions which will presently be discussed, Raunkiaer used neither the five major divisions nor the thirty minor subdivisions. He reduced the thirty to ten: (1) stem succulents, as cacti, (2) epiphytes, (3) the larger phanerophytes, or trees, (4) tall shrubs, 6 to 25 feet tall, (5) small shrubs, (6) chamaephytes, (7) hemicryptophytes, (8) geophytes, (9) hydrophytes and helophytes combined, and (10) therophytes. Next he proceeded to assign the flowering plants of the world (he never considered ferns, mosses, lichens, or fungi) to these ten classes and to determine the proportionate number of species in each class. To do this must have been an arduous task. Probably no one has ever repeated it to check on Raunkiaer's accuracy. He very likely did it by taking some large compendium of the plants of the world, as *The Natural Families of Plants* by Engler and Prantl, and leafing through it page by page, and accepting the statements of the authors as correct. There were doubtless some errors and more omissions, but we can assume that Raunkiaer remedied these to the best of his ability. His results, expressed in percentages, are as follows:

	Percent		*Percent*
Stem succulents	1	Chamaephytes	9
Epiphytes	3	Hemicryptophytes	27
Trees	6	Geophytes	3
Tall shrubs	17	Helophytes and Hydrophytes	1
Small shrubs	20	Therophytes	13

This indicates in a general way the number of species of plants in each of the ten vegetative forms and, strange as it may seem again, slight errors in his figures will not affect the conclusions which Raunkiaer reached. He called this division the *normal spectrum*, adopting the word from the physicist who passes sunlight through a prism and separates it into the seven primary colors. *Spectrum* is now commonly used in Raunkiaer's sense by botanists and we shall continue it here.

If the flora of any region of the world is known with reasonable accuracy, it is a fairly simple matter to determine its spectrum, which may then be compared with the normal spectrum given above. Raunkiaer did some of this himself; others have done it for different places, and in the following table are shown a number of spectra, taken from various sources and from

Fig. 18.3. Desert in southern New Mexico, foothills of the Dona Ana Mountains. The shrub with the long narrow leaves crowded toward the base is sotol (*Dasylirion*); the one with several long, wand-like branches from the base is ocotillo (*Fouquieria*). The leaves of sotol are thick and somewhat succulent, and resistant to desiccation. Those of ocotillo are thin and soft; they are produced only when the ground is moist, as after a rain, and when the ground dries out they fall off. (U.S. Forest Service photo by E. O. Wooton)

various parts of the world. For simplicity here, the number of classes has been reduced from ten to the original five. This has been done by combining the first five, making 47 percent for the total phanerophytes, and adding the helophytes-hydrophytes to the geophytes to make a total of 4 percent cryptophytes.

Inspection of the table shows clearly the facts which Raunkiaer wanted to demonstrate. Certain vegetative forms are represented in certain places by disproportionately large numbers of species. Phanerophytes, including trees, shrubs, succulents, and epiphytes, constitute nearly half the species of the world (47 percent), but they are below average everywhere except in the moist or wet tropics, where they are much more numerous. Less than a tenth

TABLE 2. Geographic Distribution of Raunkiaer's Life forms (*in percent*)

	Phanerophytes	*Chamae-phytes*	*Hemi-cryptophytes*	*Cryptophytes*	*Therophytes*
Moist temperate					
Alabama	18	3	48	17	14
Connecticut	15	2	49	20	14
Indiana	14	2	49	18	17
Japan	28	2	48	12	10
Germany	9	3	54	17	17
Dry temperate					
Danube Valley	7	5	55	10	23
Akron, Colorado	0	19	58	8	15
Hot desert					
Death Valley	26	7	18	7	42
Tucson, Arizona	18	11	24*		47
Tripoli	6	13	19	9	51
Cold desert					
Utah	2	23	56	5	14
Arctic					
Spitzbergen	1	22	60	15	2
Alaska (part)	0	23	61	15	1
Wet tropics					
Queensland	96	2		2	
Seychelles Islands	61	6	12	5	16
Moist tropics					
India	47	9	27	4	13
Northeastern United States,					
native plants only	17	2	49	19	13
Normal spectrum	47	9	27	4	13

* Hemicryptophytes and cryptophytes combined.

of the plants of the world are chamaephytes, but in the Arctic zone they are more numerous. The table shows more than twice the average representation, and other records, taken from still farther north, show that they may include two-thirds of the species of a region. About a quarter of all plants are hemicryptophytes, but in the temperate zone they amount to just about half. Annuals constitute 13 percent of the flora of the world, but up to 50 percent of the flora of hot deserts. In general, Raunkiaer showed that certain vegetative forms predominate in certain types of climate.

This simple conclusion could have been presented in fewer words, but it has seemed desirable to discuss vegetative form in some detail in preparation for future discussion.

Raunkiaer's data and conclusions are based solely on the number of species; they pertain to the flora of various regions. In selecting his data and in drawing his conclusions, he did not take into account three other matters which are of great interest and importance to the phytogeographer. He omitted them, not because he was ignorant of them, since they are patent to everyone, but because they lay outside his thesis. His problem was the correlation between climate and the proportionate representation of vegetative forms in the flora.

First, he took no account of the comparative rarity or abundance of some kinds of plants. For example, in the spectrum for Indiana in the table, equal weight has been given to the sourwood (*Oxydendrum arboreum*), of which only a few trees exist in the state, and the beech, which is a "frequent to common" tree in almost every county. Second, he did not consider (and there is no reason why he should have) the comparative conspicuousness of the various kinds of plants and the impression on the observer. If all the beech trees in Indiana were destroyed, it would make quite a change in the landscape; if all the little spring beauties (*Claytonia*) were pulled up, Indiana would still look about the same. Third, he did not consider the fact that there are some kinds of plants which so control or monopolize the environment that other kinds have comparatively little chance to grow. The last consideration is very important and will need more discussion presently. All of these exceptions are worth considering by the plant geographer, but it must be understood that none of them affects the validity of Raunkiaer's conclusions.

Fig. 19.1. Beach grass on sand along the Atlantic coast of the United States. (Photo by H. W. Kitchen, from National Audubon Society)

19/ The Concept of Vegetation

The term *flora*, it will be remembered, refers to the kinds of plants living together in any specified area; it does not refer in any way to their vegetative form. The term *vegetation* has an entirely different meaning and is quite independent of the flora; it refers to the general aspect of the plants of an area taken collectively and regardless of the kinds of plants which produce that aspect. It is based on the impression which the plants make on our mind through our eyesight, not individually but en masse.

We have various words in our language to designate certain types of vegetation and two of them, *forest* and *meadow*, will serve to make the meaning of the term clear. These two kinds of vegetation do not look alike; they are distinguishable by the eye at a distance. They are independent of the flora: a forest keeps the same name whether it consists of oak, pine, or maple; a meadow is still a meadow, whether the grasses are blue grass, timothy, or redtop.

In such a context the term vegetation represents a mental impression rather than a botanical phenomenon, but we can divorce our impression from the definition, get back to the botanical cause of the impression, and produce a strictly botanical definition. Why do a forest and meadow look unlike? Because each is characterized by the prevalence of plants of a certain vegetative form. Note the use of the word *prevalence*. The great bulk of the plants in a forest consist of trees, but under the trees are shrubs and herbs and even grasses belonging to different vegetative forms, but the chief impression on the mind is made by the trees. In a meadow the bulk of the plant life typically consists of grasses; various other plants of different growth form are also in the meadow, but it is the grasses which give the impression which we call meadow. Now if we omit the idea of mental impression and think only of its cause, we get the idea that the term vegetation is based on and refers to the representation of vegetative forms in an area. Different kinds of vegetation are produced by the prevalence or exclusive presence of different vegetative forms.

Exclusive presence of a single vegetative form is rare except in areas of small or minute size. The shores of Lake Superior, often composed of solid rock and cleanly washed by the waves, are then occupied only by crustose lichens; a sand dune near the Atlantic coast may be populated only by beach grass (*Ammophila arenaria*, Fig. 19.1). Such spots are usually characterized by extreme conditions of environment. Somewhat larger areas with a single vegetative form may be recognized, provided we adopt a broad classification of vegetative forms. In an American pond, all flowering plants may be referred to Raunkiaer's one general class of cryptophytes, but if we classify them among his thirty subdivisions we get quite a different result. In Table 2 the rain forest of Queensland showed 96 percent of phanerophytes, but a division of that group into subclasses would have shown a considerable percentage of epiphytes. In regions of large extent, where the same type of vegetation extends for miles, it is almost always composed of several vegetative forms, but with only one or a very few of them prevalent.

In a forest, trees constitute the bulk, the prevalent form, of the plant life. Trees tend to monopolize the water in the soil. The shade of trees greatly reduces the amount of light available to the shrubs and herbs beneath. The land supports as heavy a growth of trees as the environment permits; the other plants must make do with what light and water they can get. Trees dominate the vegetation by reason of their size and consequent control of the environment, though they may be represented by fewer species and fewer individuals. In a meadow the grasses and grasslike plants form a dense sod, with interlaced roots below the surface and a tangled mass of leaves and stems above ground. Other kinds of plants have difficulty in getting started against such competition. The herbs may be actually taller than the grasses, but grasses dominate the vegetation because of their habit. In every other kind of vegetation, there are always reasons why one or a few vegetative forms are present, but we do not always know enough to discover precisely what the reasons are.

Raunkiaer had five major types of vegetative form and divided them into thirty minor types; for his spectra of different regions he used ten types. Can every one of these five or ten or thirty types be prevalent in some favorable spot and thereby produce a distinctive type of vegetation? No, it is impossible. Epiphytes, for example, which grow perched on the branches of trees, must have other and larger plants for their seats; they can not form a vegetation of their own. Hemicryptophytes (excluding grasses), all of them herbaceous plants seldom more than six feet tall, need the same sort of environment in which trees grow well; they can not be prevalent in competition with the much larger trees. Any kind of environment which is favorable to annual plants is also adapted to various other vegetative forms. Annuals, mostly of small size, must share the space with the others, and only in deserts are they so abundant that they can be considered one of the prevalent forms. In Raunkiaer's scheme, most grasses are classified as hemicryptophytes. Their habit of growth is so different from most plants of this type that they really deserve to be classed in a separate vegetative form of their own, and they, because of their habit, can be prevalent over large areas of land.

Just as we can make a great many vegetative forms, if we wish, or reduce them to a few generalized types, as Raunkiaer did and as we did in the preceding chapter, so we can recognize a few broad and generalized types of vegetation or a larger number of minor types. The broad general types serve much better to give us a bird's-eye view of the vegetation of the world. We can classify vegetation into four such broad types. Each of these occupies some millions of square miles; each of them can be divided into minor types to as fine a degree as we wish. Each of them is composed of one or a few vegetative forms which give to each its distinctive appearance; each of them also includes other growth forms which contribute comparatively little to the general appearance of the vegetation. Besides these four, there are others of small extent and relatively small importance which are characterized by other vegetative forms. Among these latter are the vegetation of ponds, of coastal dunes, of rock cliffs, and many other areas of small size and usually of unusual environment.

The four principal types of vegetation are *forest*, *grassland*, *desert*, and *tundra*. The prevalent vegetative forms are, for the forest, trees; for the grassland, grasses; for the desert, small-leaved shrubs, succulents, and annuals; for the tundra, chamaephytes, mosses, and lichens. Each of them is distinctly correlated with a particular type of climate and each of them is developed wherever that type of climate occurs.

In the first chapter some of the difficulties of plant geography were outlined, and they have probably become increasingly evident as the plot has been unfolded. It is probably also apparent that an understanding of the subject often requires considerable thought.

Now here is something for the reader to consider:

We have found that a geographical classification can be made according to the flora, according to the plants which live together. Starting with the flora of the whole world, we can separate the flora of the Tropical Zone from those of the Temperate and Arctic. The flora of these zones can be divided into floristic groups, each occupying a province of its own, and here we reach an area with a really characteristic flora. We divide the provinces into progressively smaller areas, until we reach our own woodlot or a particular zone of plants around our own pond. Below that we can not well go. On the other hand, we can also classify areas according to their appearance, according to the vegetation which occupies them. Beginning with the vegetation of the whole world, as it might appear to an observer on Mars, we divide it first into major areas characterized by a great extent of one or the other of our four major types of vegetation. These we divide into smaller and smaller areas, each characterized by a greater similarity in vegetation, until we finally reach our own woodlot or a particular zone around our own pond. Lastly, we can disregard the plants completely and classify the surface of the world according to environment. The land is first divided into great areas having a similar general type of climate. These are in turn divided into smaller and smaller areas of progressively increasing environmental similarity, until we finally end up again with our own woodlot, which differs in some minute way from our neighbor's. We begin with the environment of the world or the plants of the world, and we end with the small plant community and its own special environment. With them we have reached the basic or unitary (not necessarily the smallest) area of plant life. The intermediate steps in the three classifications are not identical, since they are based on three different criteria. Yet all three intersect at the *province* and its *floristic group*. The province is therefore a very important area in phytogeography, since it is featured by a particular type of vegetation, its own special flora, and a characteristic type of climate.

Let us have a far-fetched illustration of this matter. One can leave Chicago on a through sleeping car on the New York Central, pass through Buffalo, and go on to New York. Also he can, or could in the past, take another sleeping car on the Grand Trunk, pass through Buffalo, and go thence to New York over the Lehigh Valley. Or he could leave Chicago on the Nickel Plate, pass through Buffalo, and go on to New York over the Lackawanna. The three routes were entirely different, yet they all started at the same place, all passed through Buffalo, and all ended at New York.

We trust that this will impress the reader with the importance of the province in phytogeography. A province is a describable area, characterized by a particular environment, its own flora, and by its own type of vegetation. It has a phytogeographical history extending many years into the past, and a future extending no one can guess how many years beyond.

20/ Correlation of Climate and Vegetation

By *climate*, the plant geographer means all those features of the environment which arise from or depend directly on the atmosphere. Of these the most important to us are temperature and rainfall. The climatologist is also deeply interested in barometric pressure, which hardly affects plants directly at all, and in the movements of great air masses. Both of these are of prime importance in determining the distribution and amount of rainfall and heat over the surface of the world, but their effect on plants is mainly indirect. Wind is, of course, an environmental factor of importance, but its immediate effect on plants is limited to increasing the loss of water, to physical damage in storms, and to a rather important role in affecting the growth of trees at timber line on mountains. Atmospheric humidity is also a matter of importance. It is closely correlated with temperature and to some extent also with rainfall, but it is also subject to great local variation depending on the nature of the terrain and the exposure. There is one place in America where it is of great importance, and that is a narrow strip of land some two hundred miles long and only a few miles wide lying along the Pacific Ocean in central and northern California. Here the air is virtually saturated with water vapor from the adjacent ocean, and the excess is often condensed into fog which in turn condenses into moisture on the vegetation. Along this narrow strip are dense forests composed of redwood (*Sequoia sempervirens*, Fig. 20.1) and several other coniferous trees, growing where the rainfall alone is entirely inadequate to support a forest. The evaporating power of the air is also important. It depends mostly on the humidity and temperature of the air, but is increased by wind. There may be occasion to refer briefly to this or some other minor climatic features of the environment, but for our present purposes we shall be limited to a consideration of temperature and rainfall in their relation to the four general types of climate.

Fortunately, climatological data have been secured from almost all parts of the world. Modern encyclopedias give statistics about the climate for many cities and countries, and most good atlases have maps showing the climate of the world. Botanists have charted the distribution of the principal types of vegetation over the world, and that is also shown on maps in many atlases. So all we have to do is compare the known climate with the known vegetation and try to find some correlation between them. We begin with the forest.

FOREST CLIMATE

Forests occupy more than half of the tropical zone. In the Warm Temperate Zone just south and north of the tropics there are also forests, but they are exceeded in area by deserts and grassland. In the Cold Temperate Zone, which we define as land where one month or more of the winter has an average temperature below the freezing point, and which occupies, in general terms, the northern half of the United States and southern Canada, forests are also extensive and probably occupy more than half of the land, although there are huge areas of desert in the interior of Asia. Forests extend north to the arctic tree line, which fairly well approximates the Arctic Circle, sometimes passing beyond it, as in Norway, sometimes falling short of it, as in Labrador. Always the dominant vegetative form is the tree, and all trees, no matter what their kinds or their habits, share one important feature: the living parts of their stems, branches, and twigs, and often also their leaves, stand exposed to the elements the whole year around. That is the reason Raunkiaer invented the term *phanerophytes*, meaning exposed plants. Obviously trees must be able to live without being destroyed or seriously injured by this exposure. Since trees do not live everywhere in either the Tropical or the Temperate Zone, there must be something in the environment of the treeless areas to prevent them. What is it?

We think at once of three possibilities: extremes of heat, extremes of cold, and insufficient moisture. So far as heat is concerned, we can find no place in the world, except possibly small areas very close to volcanic craters, where the climate is too hot for trees. Look at our own American deserts with their irrigated orchards of oranges and grapefruit and their great trees of eucalyptus (Fig. 20.6); remember the hot oases of the Sahara with their groves of tall date palms. These are among the hottest places in the world, with temperatures exceeding 120 degrees, yet trees are able to live there when provided with sufficient water. As to extreme cold, think of the trees growing along the Yukon River in Alaska, where the thermometer frequently drops to −50 degrees. Even in Siberia, where the temperatures reach 70 below zero, trees still grow. It seems that when temperatures drop below zero, additional cold makes little or no difference to many kinds of trees.

Fig. 20.1. Redwood forest in California, near Crescent City. The luxuriant undergrowth is typical. (U.S. Forest Service photo by W. I. Hutchinson)

Naturally it is these cold-resistant kinds of trees which can and do live farthest north. The northern boundary of the forests is set not by the minimum temperature of the winter, but by conditions of the summer. Trees, like all other plants, require a considerable degree of warmth, continued for a reasonable time, for the completion of their annual life processes of growth, flowering, ripening of seeds, and production of buds for the following year. It is probable that the amount of heat necessary and the length of the summer season during which this heat is required will vary for different kinds of forest trees. Certainly the number of kinds of trees decreases gradually from the tropics northward until only a very few kinds are left to mark the northernmost outposts of forest. The best we can do is to get some idea, from the scanty climatological records of the far north, of the length and temperature of the arctic summers. From these it has been generalized that trees must have a growing season at least eight weeks long with a temperature averaging 50 degrees or more. If the summers are either shorter or colder, forests can not exist.

Toward the south, a tree limit is not reached in South Africa, Australia, or New Zealand. South America extends much farther south, but it tapers to such a narrow point that the influence of the ocean keeps the winters relatively warm and the summers relatively cold. Tierra del Fuego, lying between the Straits of Magellan and Cape Horn, has a fairly long but very cold summer with an average temperature of 50 degrees, and is partially covered by forest. In the Falkland Islands the warmest two months average only 47 degrees, and the islands are treeless. The so-called Antarctic Islands of the South Indian Ocean have relatively warm winters and cold summers and are treeless. Some of these islands may lack trees merely because of their isolation. Tristan da Cunha, in latitude 37° south, has just one species of tree, although the climate appears to be favorable for trees.

Similarly, the location of the tree line (popularly called the timber line) on mountains in the Temperate Zone is set primarily by the length and warmth of the summer. In general terms, each additional 330 feet of altitude causes a drop of one degree in temperature, and this holds pretty well for both winter and summer. So far as the effect of temperature alone is concerned, the figures given for the arctic tree line will serve in a general way for mountains in the Temperate Zone, that is, a growing season at least 8 weeks long with an average temperature of 50 degrees or more. But the situation on mountains is complicated by two other factors, snowfall and wind. Snowfall is heavy in some American mountains, notably the Sierra Nevada and many of the lower mountains of New England. Snow on these mountains tends to lie long on

Fig. 20.2. Timber line in the mountains of western United States. Top, in New Mexico, with Engelmann spruce. Bottom, in Colorado, with willow in right foreground, spruce and fir in left background. (U.S. Forest Service photos; top by Rex King, bottom by H. E. Schwann)

the ground, preventing all plant growth merely by its presence, and taking so long to melt that the effective length of the summer is considerably reduced. Even below timber line, there are spots of limited size where snow tends to accumulate in drifts, and these may be so slow in melting that the establishment of tree seedlings is impossible. Such places are seen by the summer tourist as open glades surrounded by forest.

The effect of wind on mountains is not so much to prevent the growth of trees and thereby lower the timber line (Fig. 20.2) as it is to dwarf them and reduce them, at or near their upper limit, to gnarled, crooked, picturesque shapes entirely unlike their normal growth habit. This type of forest, commonly called wind-timber or more poetically elfin forest, is beautifully exhibited on many readily accessible peaks in the Rocky Mountains.

An interesting effect which is apparently due to exposure to wind is common in tropical mountains, where the upper limits of the forest are marked by an extensive development of dwarf (up to twenty feet; usually not more than ten feet), widely branched, gnarled and crooked trees, growing in a dense, tangled, impenetrable mass. Such growth is often called mossy forest (Fig. 20.3), because the branches, twigs, and even the leaves are heavily covered with mosses and liverworts. The mossy forest is relatively independent of temperature. In the Sierra de Luquillo of Puerto Rico it may occur as low as 2,300 feet, and in the Andes it may still exist at more than 10,000 feet. Always it represents the last outpost of the forest. It extends to and over the summits if they are not too high, or gives way to the barren land of the higher mountains. Always it is correlated with very moist atmosphere and excessive exposure to wind. There is nothing like it within the United States; our nearest approach to it is on the "balds" (Fig. 22.19) which cover the tops of some of the higher mountains in the southern Appalachians. There a jungle of shrubs is composed largely of azalea and rhododendron and is a gorgeous sight during the blooming season. The cause of these treeless summits is unknown, but exposure to wind may have a great deal to do with it.

The tree line on high tropical mountains is an entirely different phenomenon from that in the Temperate Zone. As stated above, the length of the growing season is the most important factor in this country. In the tropics there is no difference, or very little difference, in the mean daily temperature throughout the year. In Java, for example, the difference between the warmest and the coldest months is only 2 degrees. Farther from the equator, either north or south, the difference increases. At Havana, Cuba, it amounts to 10 degrees. The difference is 15 degrees at Key West, Florida, which lies just within the

Fig. 20.3. Mixed hardwood forest in Puerto Rico. The dark forest on the ridges in the background is of the dwarf, mossy forest type. (U.S. Forest Service photo)

Temperate Zone on the map but has a tropical flora. The variation from day to night is usually more than the change from season to season. At progressively higher elevations, accompanied by progressively lower average daily temperatures, an altitude is eventually reached at which frost sometimes occurs at night, while at still higher levels frost occurs every night and lasts for longer hours. Many kinds of tropical plants can not endure frost, and these plants of tropical ancestry and behavior are nipped off one by one with increasing altitude. Similarly, with increasing altitude the warm period of the daytime is shortened and reduced in temperature. It used to be postulated that plants need a temperature of 43 degrees or more for their successful growth. If we can for the moment accept this perhaps oversimplified postulate, we can imagine that various kinds of plants, not injured by frosty nights, are unable to live at the high altitudes for lack of sufficiently high temperature over a long enough period each day. As a matter of fact, we do not *know* the cause of the tropical tree line. We have merely suggested two factors which may, and probably do, play some part in determining it.

Between the arctic and antarctic extremes, where the tree line is determined primarily by length of season, and below the tree line on mountains, desiccation becomes a more important factor than temperature in the life of a tree. The living parts of all plants contain much water, and most plants are promptly killed by excessive loss of water. (There are exceptions, of course. Some ferns and many mosses and lichens can be dried until they will crumble in the fingers and still resume their vital processes when moistened, but we are now talking about trees.) The buds of our trees in winter, the leaves in summer, and the leaves of evergreen trees at all seasons are constantly subject to loss of water. The danger from this to the life of the plant can be reduced in three principal ways: by protection of the exposed parts with a waterproof covering, by the availability of a continuous supply of replacement water from the soil, and by growth of the plant in a saturated or nearly saturated atmosphere in which evaporation is reduced.

Trees make use of all three methods. The surface of leaves and young twigs is covered by a thin but essentially waterproof coating called the cuticle, but it is punctured by myriads of minute openings, the stomata, through which gases such as oxygen and carbon dioxide are exchanged with the atmosphere and through which water vapor also escapes. The bark which covers the trunk, branches, and twigs is composed partly of cork. This greatly reduces loss of water from the stems but is by no means perfect. The greatest loss of water is through the leaves, and most broad-leaved trees drop their leaves in winter, thereby greatly reducing the loss of water. A good supply of soil water is of no use to the tree or to any other kind of plant if the soil is frozen, nor can it be conducted up the trunk if the trunk is much below the freezing point. Soil water is readily available to plants at all seasons in warm climates but only during the summer in cold climates. The effectiveness of cuticle and cork in preventing loss of water needs to be supplemented, and that is best done by a moist atmosphere in which water evaporates slowly. Moist air, as a prevailing or common condition, is usually associated with a climate with plenty of rain. Obvious exceptions may occur to you, but we are trying to present a general picture along the broadest possible lines. One exception has already been mentioned in the first paragraph of this chapter; a second will be discussed later.

The general conclusion to which we are led is that forest, as the prevailing type of vegetation, is characteristic of regions with *adequate* rainfall at all times of the year.

What can be called an adequate rainfall? It is impossible to reason this out, but we can arrive at a very good general idea by studying the average rainfall at a considerable number of places where forests grow. Also, if we can find out what the rainfall is in places located at or near the edge of the forest, where it disappears in favor of some other type of vegetation, we can learn what the minimum requirements of the forest are. So we find that in the cold Temperate Zone, with one to several months of each year averaging below the freezing point, the rainfall must be 10 or better 15 inches per year at the north, increasing toward the south. In the warm Temperate Zone, where the coldest month is above the freezing point, it should be 35 inches or more, and in the tropics at least 60 inches. There are exceptions to these figures, as to almost every statement in plant geography, but in general they give a reasonable idea of the matter.

These are stated as the approximate minimum amounts of rainfall for a forest. Greater rainfall does not produce a different type of vegetation. The individual trees are larger and the forest growth is denser, so that the total bulk of vegetation is larger, and there may be more different kinds of trees, but it is still a forest.

In addition to these requisite amounts of rainfall, there must be a proper seasonal distribution of the rain, since forest trees require moisture at all times of the year. In considering conditions for the Temperate Zone, we may distinguish a growing or summer season from April to September and a dormant or winter season from October to March. For the tropics, where temperatures are essentially uniform throughout the year, we may often distinguish dry and wet seasons. This is particularly characteristic of India, where the southeast monsoon blows for about six months and brings heavy rain to almost all parts of the country, to be followed by the northeast monsoon bringing six months of drought. In general, each half of the year should receive at least 35 percent and preferably 40 percent of the total amount if the total is at or near the minimal requirements. That means a minimum of 4 to 6 inches in each half of the year in the northern part of the cold Temperate Zone, increasing southward, of 12 to 15 inches in the warm Temperate Zone, again increasing southward, and of 20 to 25 inches in the tropics. If one season has excessive rainfall, then the other may have the minimal amount, which may be much less in proportion. To illustrate: the heaviest growth of forest in the world is found in western Oregon and western Washington, where the summer rainfall is only about 20 percent of the total. The actual summer rainfall, however, amounts to 20 inches or more, which is well above the required minimum.

The United States is a good place to check the validity of these figures, since it extends from well into the cold Temperate Zone along the Canadian boundary to far into the warm Temperate Zone along the Gulf Coast and in the southwest. The eastern half of the United States is or was almost wholly forested, but much of the interior, west of the forest, is occupied by grassland or desert. The tables which follow show the rainfall, both total and seasonal, along several east-west lines, beginning where the country is well forested and extending westward into treeless country. The information is taken from publications of the United States Weather Bureau. Stations of the Weather Bureau were not located especially for the benefit of plant geographers, nor are they spaced at equal distances from each other. The statistics nevertheless show this rather intimate relation between amount and distribution of rainfall, temperature of the dormant (winter) season, and distribution of forests.

In tropical lands the rainfall is often great, exceeding 100 inches per year in many places and in some reaching 200 or even 300 inches. Several types of forest occur in the tropics, differing in their density, in the size of the trees, in the nature of the accessory plants which live with the trees, and to some extent even in the habits of the trees themselves. It is beyond the scope of this book to discuss the climatic requirements of the various types, but it may be stated that regions with 60 inches of rain are usually well forested, and that 80 inches or more, if distributed evenly throughout the year, will support a rain forest, which is the highest expression of tropical forest development.

Fig. 20.4. Grassland in Nebraska. (U.S. Forest Service photo by Bluford W. Muir)

TABLE 3. Climatic Changes

	*Rain in Winter**	*Rain in Summer**	*Total Rain*	*Months below Freezing**	*Temperature of 3 Warmest Months*	*Longitude*	*Vegetation*
APPROXIMATELY ALONG LATITUDE 47 DEGREES NORTH							
Sault Ste. Marie, Mich.	12.2	13.7	25.9	4	61	84° 21′	Forest
Duluth, Minn.	9.0	20.9	29.9	5	63	92° 06′	Forest
Park Rapids, Minn.	6.4	20.5	26.9	5	65	95° 10′	Forest
Moorhead, Minn.	6.4	18.1	24.5	5	66	96° 44′	Grassland
Grand Forks, N.D.	4.4	15.7	20.1	5	65	97° 05′	Grassland
Oakdale, N.D.	4.7	13.2	17.9	5	65	102° 50′	Grassland
Glendive, Mont.	5.2	10.7	15.9	5	69	104° 30′	Grassland
Great Falls, Mont.	3.5	9.9	13.4	3	66	111° 15′	Grassland
APPROXIMATELY ALONG LATITUDE 43 DEGREES NORTH							
Port Huron, Mich.	14.3	16.7	31.0	4	66	82° 26′	Forest
Grand Haven, Mich.	16.0	18.8	34.8	4	67	86° 13′	Forest
Milwaukee, Wis.	12.6	18.4	31.0	4	68	87° 54′	Forest
La Crosse, Wis.	9.2	21.7	30.9	4	71	91° 15′	Forest
Charles City, Iowa	8.2	21.6	29.8	4	71	92° 40′	Forest
Sioux City, Iowa	5.6	19.9	25.5	4	72	96° 24′	Grassland
Rosebud, S.D.	5.5	12.8	18.3	4	71	100° 52′	Grassland
Oelrichs, S.D.	7.2	12.1	19.3	4	70	103° 13′	Grassland
Fort Laramie, Wyo.	4.5	9.7	14.2	3	71	104° 31′	Grassland
APPROXIMATELY ALONG LATITUDE 40 DEGREES NORTH							
Pittsburgh, Pa.	16.5	20.3	36.8	1	73	80° 02′	Forest
Columbus, Ohio	18.7	19.5	38.2	1	73	83° 00′	Forest
Indianapolis, Ind.	19.4	22.5	41.9	1	74	86° 10′	Forest
Springfield, Ill.	16.1	21.3	37.4	2	74	89° 39′	Forest
Keokuk, Iowa	12.3	22.8	35.1	3	75	91° 26′	Forest
Corydon, Iowa	10.5	24.8	35.3	3	73	92° 40′	Forest
Atchison, Kans.	10.1	27.0	37.1	3	75	95° 08′	Forest
Concordia, Kans.	6.5	20.3	26.8	2	76	97° 41′	Grassland
Colby, Kans.	3.6	15.2	18.8	3	74	101° 02′	Grassland
Cope, Colo.	3.6	14.8	18.4	3	72	102° 49′	Grassland
APPROXIMATELY ALONG LATITUDE 37 DEGREES NORTH							
Cairo, Ill.	21.2	20.4	41.6	0	77	89° 10′	Forest
Olden, Mo.	18.8	22.6	41.4	0	75	91° 54′	Forest
Springfield, Mo.	17.2	26.4	43.6	0	75	93° 18′	Forest
Columbus, Kans.	14.8	29.7	44.5	0	77	94° 50′	Forest
Independence, Kans.	12.5	24.5	37.0	0	79	95° 43′	Grassland
Englewood, Kans.	5.1	15.4	20.5	0	78	99° 58′	Grassland
Viroqua, Kans.	4.2	13.5	17.7	0	76	101° 46′	Grassland
Blaine, Colo.	3.6	11.7	15.3	1	75	102° 15′	Grassland

TABLE 3. Climatic Changes (*concluded*)

	*Rain in Winter**	*Rain in Summer**	*Total Rain*	*Months below Freezing**	*Temperature of 3 Warmest Months*	*Longitude*	*Vegetation*
APPROXIMATELY ALONG LATITUDE 34 DEGREES NORTH							
Memphis, Tenn.	23.4	27.4	50.8	0	80	90° 03′	Forest
Little Rock, Ark.	25.5	24.1	49.6	0	79	92° 06′	Forest
Fort Smith, Ark.	18.5	23.3	41.8	0	79	94° 24′	Forest
Oklahoma City, Okla.	11.3	20.4	31.7	0	78	97° 33′	Grassland
Fort Sill, Okla.	10.3	19.8	30.1	0	80	98° 25′	Grassland
Roswell, N.M.	4.4	11.2	15.6	0	77	104° 30′	Grassland
APPROXIMATELY ALONG LATITUDE 30 DEGREES NORTH							
New Orleans, La.	25.6	32.0	57.6	0	82	90° 04′	Forest
Lake Charles, La.	26.9	26.4	53.3	0	81	93° 06′	Forest
Houston, Tex.	21.2	27.0	48.2	0	82	95° 15′	Forest
San Antonio, Tex.	10.7	17.7	28.4	0	82	98° 28′	Scrub
Fort Clark, Tex.	7.4	16.0	23.4	0	84	100° 24′	Desert
El Paso, Tex.	3.0	6.3	9.3	0	80	106° 30′	Desert
Tucson, Ariz.	4.2	5.6	9.8	0	85	110° 53′	Desert

* In this table, "winter" refers to the six coldest months of the year, and "summer" to the six warmest months; "rain" is total precipitation (both rain and snow), measured in inches; temperature is given in degrees Fahrenheit; and "months below freezing" refers to months in which the mean temperature is below 32 degrees Fahrenheit.

GRASSLAND CLIMATE

The habits of sod-forming grasses have already been described but may be mentioned here again. All winter buds of grasses lie dormant near or below the surface of the ground, well protected against excessive loss of water by the dead litter of last year's leaves. They have little need for water through the whole of this dormant period. Spring arrives, the buds develop, new leaves are produced on short stems, and a thick green carpet develops. Late spring and summer see longer stems rising above the sod, bearing leaves at longer intervals, and bearing flowers (mostly terminal) and later seeds. During this growing season the grasses need an adequate supply of water. Late summer and autumn arrive, the seeds are ripe, the tall stems turn brown and dry, the leaves also wither and dry, and the year's work is done. Autumn rains are of little value to the grasslands.

There is nothing in the preceding paragraph to indicate that grasses could not grow well with heavier winter rain. As a matter of fact, most grasses can grow well in a forest climate if given the proper opportunity. Most grasses, however, demand plenty of light and do not grow well under the forest shade, where they are reduced in number of individuals and do not form a sod. In competition with the forest they are accordingly at a great disadvantage. Grassland climate must therefore be unadapted to forest as well as adapted to grasses. Typically that means low winter and adequate summer rainfall.

As to temperature, grasses are no more demanding than forests. Our American grasslands extend far north into Canada and are well developed in places where temperatures of 50 and even 60 degrees below zero have been recorded. They extend south into the warm Temperate Zone, into regions with long hot summers and merely cool winters; they reappear in the tropics where the temperature varies but little throughout the year.

The rainfall needs of our American grasslands can be seen from the preceding tables, which show the climate along 5 strips in different latitudes. Inspection of these tables will show a distinct decrease in winter rainfall (except at Latitude 47) between the forest and grassland, and usually a drop in summer rainfall as well.

Another feature of grassland climate not evident from the tables is the frequency of summer drought on the grassland. Droughts also occur in the eastern forests as far as the Atlantic coast, but not so frequently. Droughts of one to eight weeks, with a rainfall a fifth of the normal or less, are a common feature of the grassland. Since the normal rainfall of the grassland is already below that of the forest, a subnormal rainfall must be of considerable importance in setting a western limit to the forest.

Extensive grasslands also occur in the tropics, especially in Africa and Australia, but also in America, as the campos of Brazil and the llanos of Venezuela. They are always correlated with a climate marked by strongly differentiated wet and dry seasons. Two examples will be sufficient:

Loango, West Africa:	dry season, June to Oct.	1 inch
	wet season, Nov. to May	58 inches
Minas Geraes, Brazil:	dry season, April to Sept.	11 inches
	wet season, Oct. to March	51 inches

In the south Temperate Zone, grassland is well developed in South Africa, Australia, and South America (the pampas):

Natal, South Africa:	dry season	4.6 inches
	wet season	23.2 inches
Corrientes, Argentina:	dry season	15.2 inches
	wet season	30.8 inches

TUNDRA CLIMATE

North of the northernmost forests of North America and Eurasia, and south of the continents of the southern hemisphere in the outlying islands, the landscape is devoid of trees. The northern areas, which are by far the more extensive, are characterized by long cold winters, averaging not far from zero and often still colder, and by summers which are too short or too cold to support trees. In the southern hemisphere, where the surrounding ocean keeps the temperatures fairly equable throughout the year, the Antarctic Islands are characterized by the same cold summers, although the winter temperatures are sometimes surprisingly high in comparison. The coldest two months in the treeless Falkland Islands, for example, have about the same average temperature as the coldest two months in Washington, D.C.

Since the tree line is apparently determined by the length and temperature of the summers, the southern boundary of the tundra might at first thought be defined merely by the absence of trees and characterized by a climate too cold for trees. That is not quite correct, for tundra extends well to the south of the theoretical tree line, not everywhere, but only in certain habitats, into areas warm enough for trees but unfitted for tree growth by other local conditions.

Throughout the cold Temperate Zone the surface layers of soil are apt to freeze during the winter. In the warmer parts of this zone, let us say Columbus, Ohio, or New York City, the ground may be frozen only for short periods, thawing between times, and the frost penetrates only a few inches into the soil. Home-owning readers will know that their water pipes must be buried deep enough to escape danger of freezing, and this depth is often prescribed by law after being learned by experience. With longer and more intense periods of cold, as at Detroit or Albany, the frost penetrates deeper and lasts longer. A period of warm weather in the winter may thaw the surface for a few inches, leaving a layer of frozen soil deeper down. The home gardener also knows that the surface of the ground remains very soggy when frozen soil persists beneath it, because the natural drainage is greatly impeded by the frozen subsoil. Still father north, as at Montreal or Winnipeg, the frost remains in the subsoil all winter and persists longer in the spring.

In the Arctic this condition is still more pronounced, and on the tundra (Fig. 22.1) the frozen soil actually lasts all summer, producing the phenomenon known as permafrost. This layer of hard, impervious, permanently frozen soil underlies all of the tundra and extends down to great depths, depending on the average mean temperature of the winter. In Siberia it has been reported to reach a depth of 2,000 feet—whether it does or not is of small consequence to the vegetation, for a thin layer will be just as effective as a thick one in stopping all root growth and interfering with natural drainage. When the short Arctic summer arrives, the surface of the soil begins to thaw, and the thawing proceeds to a depth that depends on the average temperature and length of the summer. Thawing produces a superficial layer of soil permanently soggy with water and usually peaty in nature, which restricts plant life to those species which can stand a cold, saturated soil. If the depth of the soil is sufficient and the temperatures are right, even trees can grow over permafrost, and it is noteworthy that the two commonest trees along the Arctic tree line in Canada are the bog-loving tamarack (Fig. 14.2) and black spruce (Fig. 14.1).

Rainfall in the Arctic is always low, rarely more than 10 inches (including that which falls as snow), and often less than 5. Such an amount could support only a desert in warmer climates. In the Arctic, low temperatures reduce evaporation, and permafrost interferes with drainage, so that W. C. Steere, botanical explorer of arctic America, has aptly if facetiously called the tundra the "wettest desert in the world."

The low altitude of the sun above the horizon at high latitudes lowers the intensity of the sunlight, but that is largely compensated by the very long days in summer. There is no good evidence that any Arctic plants prefer that region, or that any other plants are excluded from the Arctic, because of the amount of available light or the length of the summer days, although it is true that some kinds of non-Arctic plants, such as the cultivated tomato, die in continuous light under experimental conditions.

There is also a northern limit to arctic vegetation. With each mile farther northward, the summers become colder and shorter and the thawed soil over the permafrost becomes thinner and is available for a shorter time. One by one the Arctic species of plants disappear, unable to grow and propagate in the progressively shorter seasons, until finally the last of them has vanished and the land is left to bare rocks, ice, and snow.

In summary, the arctic tundra is characterized by short cold summers and by a thin layer of available soil overlying permafrost. Its northern limit is set by increasing shortness of the growing season and thinness of soil, and its southern limit by a soil layer thick enough and a summer long enough to support trees.

DESERT CLIMATE

It is not so easy to define a desert as one might at first think. Webster says it is an arid region in which the vegetation is especially adapted to scanty rainfall.

Fig. 20.5. Desert in southern Nevada. Joshua tree (*Yucca brevifolia*) in right foreground. (U.S. Forest Service photo by Paul S. Bieler)

That is good, but it leaves us to define the word scanty. Surely we can not say that the rainfall is too scanty for the plants which habitually live in the desert; there are hundreds of kinds which live there successfully, and many of them can scarcely be coaxed to live anywhere else. In most deserts a considerable proportion of the surface is not occupied by plants. Along an irrigation ditch, where some water escapes by seepage into the desert, there are more individual plants and they reach a larger size. In a desert, then, the ordinary supply of water, coming from rainfall, is not sufficient to produce the maximum bulk of vegetation. Water is the limiting factor, just as light is the limiting factor which restricts the number of individual herbs beneath an ordinary forest. A desert is therefore a region in which the water supply, so far as it depends directly on rainfall, is at or near the minimum at which plants may live. There are a few areas, mostly of limited extent, where no rain falls at all, at least not for several successive years, and which are completely barren. Best known of these to Americans is the Atacama Desert of northern Chile, but similar areas are said to occur in the Sahara, Arabia, and central Australia.

All deserts are hot, at least at some season of the year. Even the northernmost ones, like the Gobi of Mongolia, have hot summers, although they may be bitterly cold in winter. Scanty rainfall is correlated with cloudless skies and intense sunshine, and it is not surprising that the highest natural temperatures in the world are always recorded from deserts. In southeastern California and southwestern Arizona readings above 120 degrees have often been taken, and 130 degrees has been exceeded both there and in Algeria.

In general, the amount of potential evaporation from plants increases with the temperature. In our northern deserts, where there is a fairly long dormant season in the winter and a comparatively short and cool summer, less rainfall is needed to support plant life. Forests in the same latitude succeed with only 15 inches per year, and the deserts obviously have considerably less. Farther south in our country, where the winters are short and mild and the summers long and remarkably hot, deserts develop even with considerable amounts of rain, amounts which would maintain a good forest in a cooler climate. In general we may say that along the equator (and deserts are scarcely found there) a rainfall of 40 inches or less results in desert. In the subtropical zones, where the largest deserts are located, the rainfall of deserts is 20 inches or usually much less; in cold temperate zones, a rainfall of less than 10 inches ordinarily results in desert.

There follow illustrations of the climate at six stations in American deserts. Three of them may properly be called hot deserts and are located in the

Sonoran Province; the other three are cold deserts in or near the Great Basin Province.

	Tempe, Ariz.	*Imperial, Calif.*	*Needles, Calif.*	*Deseret, Utah*	*Benton City, Wash.*	*Ephrata, Wash.*
Latitude	33° 26′	32° 51′	34° 50′	39° 18′	46° 15′	47° 19′
Altitude (*above sea level*)	1159	−69	471	4541	680	1275
Average annual rainfall (*in inches*)	8.10	3.24	5.17	6.57	8.60	8.18
Coldest day on record (*in degrees Fahrenheit*)	15	22	21	−32	−29	−23
Average January day (*in degrees Fahrenheit*)						
Minimum	34	37	39	12	23	18
Maximum	64	71	64	38	38	33
Hottest day on record (*in degrees Fahrenheit*)	113	124	125	106	114	112
Average July day (*in degrees Fahrenheit*)						
Minimum	73	76	80	55	56	61
Maximum	104	107	108	93	93	92

ANOTHER TYPE OF FOREST AND ITS CLIMATE

It is sometimes convenient to have a simple aid to memory by which a series of facts may be recalled. One of the most familiar of these is the word *vibgyor*, by which the colors of the spectrum may be remembered in their proper order. We can make an analogous system for remembering the correlation between rainfall and vegetation in the temperate zone. We have to omit the arctic tundra, because there the correlation is between vegetation and temperature, and the Tropical zone, where there is little or no difference between the summer and winter seasons.

Suppose we indicate adequate winter rainfall for a forest by a capital W, and adequate summer rainfall by a capital S. Then winter rainfall too low for a forest can be indicated by a lower-case w, and deficient summer rainfall by a lower-case s. The formulas to be remembered are then

SW—Forest Sw—Grassland sw—Desert

Immediately it will be noticed that there is another possible combination which does not appear in the formulas above: sW, with a low summer and high (or relatively high) winter rainfall. Is there anywhere in the world a climate of that sort, and if so, what kind of vegetation occupies it?

Such a type of climate exists only in the Warm Temperate Zone just north of the belt of deserts in the northern hemisphere and just south of the deserts in the southern hemisphere, and then only on the western side of continents, where it may extend inland for many miles. We think at once of the great Sahara Desert of northern Africa. Just north of it lies the Mediterranean Sea, and the lands along its shores have precisely this type of climate, which takes its name from this location and is generally called the Mediterranean climate. Travelers to Spain or Italy in winter may remember the fairly mild temperatures and frequent rains; summer visitors will remember the rainless days, the cloudless skies, and the heat. On the western side of North America, just north of the deserts of northwestern Mexico, the same kind of climate appears in California and characterizes the Californian Province. In the Southern Hemisphere the same climate appears around Cape Town on the west shore of extreme South Africa and just south of the South African deserts; it appears again in southwestern Australia, just south of the great Australian deserts, and extends eastward in a narrow strip for many miles along the south shore of Australia. It also occurs in Chile on the west coast of South America just south of the Chilean deserts.

The greatest extent of land with this type of climate is around the Mediterranean Sea, where it includes most of Spain, Italy, and Greece, and the coast of Turkey and Israel, and some of the parts of northern Africa which project farthest to the north. This region is the home of the olive, the fig, and the vine. These well-known plants, as well as others, are commonly planted in each of the four other areas of Mediterranean climate which we have mentioned. Eastern visitors to California are always impressed by the groves of olive and fig and the enormous vineyards. Also they will find in California tall trees of eucalyptus (Fig. 20.6) from the corresponding region of Australia, and from South Africa occasional silver trees and numerous plants of the iris family.

In each of these five regions the climate is about the same. There is a long hot summer which is rainless or nearly so, and a short, mild, rainy winter, during which frosts seldom occur. If we divide the year into the customary two halves, as has been done before in this chapter, we find that only 10 to 20 percent of the rain falls in the warmer six months. Much of this little rainfall comes in April (October in the southern hemisphere, where the seasons are reversed) and may be regarded as a continuation or conclusion of the winter rains. So, if we place April (or October) into the winter half of the year and October (or April) into the summer half, we find that the summer rainfall is often only 5 to 10 percent of the total.

Fig. 20.6. Eucalyptus (*Eucalyptus globulus*) in California. The climate of California is similar to that of the native home of eucalyptus in Australia, and the trees do well here. (U.S. Forest Service photo by H. D. Tiemann)

How can a forest exist through such a protracted drought? There are three answers to this question. First, the kinds of forest trees which have this ability are comparatively few; in California they do not exceed twenty. Second, the forest is never extensively developed over the whole area with such a climate. No exact figures are available, but it is probable that the forests of the Californian Province did not occupy more than a tenth of the total area. Most of the vegetation is a dense growth of shrubs, and the forest is commonly restricted to spots where soil water tends to accumulate during the rainy winter, such as the bottom of valleys, or to the sides of certain hills where the slope of the underlying strata of rock tends to deliver soil water to the surface, or to low mountains which catch a larger rainfall. And third, the trees themselves are built in a different way from those of ordinary forests. In most of them the leaves are thick, stiff, hard, and evergreen, and their stomata, the minute openings in their surface through which water evaporates, are closed during most of the summer. The trees actually go into a period of semi-dormancy during the dry season. The same type of structure is typical for the shrubs also.

The greatest number of species of trees, the best development of forest, and the largest area of forest of this type are in southwestern Australia, where several species of eucalyptus form dense forests and tower to heights of 200 feet or more. Here the rainfall is from 30 to 60 inches. A long strip of land extending a thousand miles east along the south shore of Australia has a similar climate but with less rainfall; here the vegetation is composed largely of shrubby species of eucalyptus known as mallee. Forests in the Mediterranean region were never extensively developed in historic times and were mostly restricted to the mountains where the rainfall is greater.

	Perth, Australia	*Cape Town, South Africa*	*Santiago, Chile*	*Malaga, Spain*	*Auberry, Calif.*
Total rainfall (*in inches*)	36.0	24.9	16.6	23.9	26.0
Rainfall, winter half (*in inches*)	31.2	19.4	15.1	18.4	22.7
Rainfall, summer half (*in inches*)	4.8	5.5	1.5	5.5	3.3
Average temperature, coldest month (*in degrees Fahrenheit*)	55	57	49		43
Average temperature warmest month (*in degrees Fahrenheit*)	74	66	62		79

To forests of this type the German phytogeographers gave the name Hartlaubwald, which means literally hard-leaf-forest, and a very appropriate term it is. Of course, as we all know, most technical terms in English are derived from Greek or Latin. So English-speaking botanists have taken two Greek words, *scleros* meaning hard and *phyllon* meaning leaf, and combined them to make the term sclerophyllous forest, which is equally appropriate and far more impressive and mouth-filling.

So now we shall write the fourth and last member of our formulas as

sW—Sclerophyllous Forest.

Some climatic data for five sclerophyllous forest areas, on as many continents, are given in the table at the bottom of the opposite page.

Fig. 21.1. Gallery forest of cottonwood (*Populus deltoides* var. *occidentalis*) in Nebraska. (U.S. Forest Service photo by Carl A. Taylor)

21/ On Transitions between Provinces

The phytogeographer can seldom propose a general principle to which exceptions can not be found. We have just tried to show the general correlation between climate and vegetation in the temperate zone, and, *in general*, everything that has been said is true. But here again there are exceptions and variations and deviations. Fortunately for us, they are not serious; in fact, they look much more formidable when committed to paper than when encountered in our travels. As a preparation for such possible encounters, we had better mention a few of them and thereby prepare the student for others as well.

The Eastern Deciduous Forest Province is characterized in its climate by plenty of rain, distributed fairly evenly throughout the year. It is therefore occupied by forest, as its name indicates. But, and here is where the exceptions occur, not all of the area has an adequate water supply, notwithstanding the ample rainfall. There are rock cliffs with very little soil to hold water, rock ledges where water runs off immediately, and sand dunes with little capacity for holding water and with unusually good drainage. Such spots obviously can not support a forest. Their plants are mostly small and widely scattered, often limited to a few species, and in some respects they are suggestive of deserts. Do not be misled into calling them deserts; they are utterly unlike the deserts of Utah or Arizona. In every one of them there is a constant trend toward a forest. On the rock ledges and cliffs, frosts disintegrate the rock physically; mosses and lichens disintegrate it chemically; soil accumulates in pockets and crevices and is protected from washing away by a covering of litter. Larger plants appear and leaf mold accumulates more rapidly, until finally, after hundreds or even thousands of years, these bare rocks are covered by forest. Sand dunes, either bare or with sparse vegetation, look very much alike, whether on the shore of the Atlantic or the Great Lakes, or inland in a real desert, yet plants are constantly at work on them, gradually stabilizing the sand by roots beneath the surface and by an accumulation of litter above, adding humus to the sand, and preparing the dunes for a denser vegetation which finally culminates in forest. There is no better place to observe the progressive development of dune vegetation than on the shores of Lake Michigan, where in a half-day excursion one may observe every stage from migrating dunes devoid of plants to completely stabilized dunes occupied by dense forest. Similarly, in the northern half of the United States are thousands of small lakes where the plant life is restricted to aquatic plants. The history of such lakes has already been discussed: vegetable matter formed by the plants themselves, and silt carried in by the affluent streams gradually fill up the lake, which is eventually occupied by forest.

Yes, the culminating type of vegetation throughout the Eastern Deciduous Forest Province is forest. These apparent exceptions, and others which we can easily find, are merely retarded stages in the development of forest. Time is the important factor, and we should never forget the importance of time in plant geography. Similar apparent exceptions can be found in every other province, and almost every one of them can easily be explained in some similar way.

One of the more striking and widespread of these anomalies is to be seen in the Middle West. It would have been much more conspicuous if we could have seen it a century and a half ago. We would then have seen great areas of grassland and somewhat smaller areas of forest growing side by side in the same climate and apparently equally well developed. An explanation of this condition has already been given in Chapter 17 and need not be repeated here.

Even beyond the limits of a forest climate, strips of forest follow the rivers of Kansas and Nebraska, where they are easily observed from various railways and highways. There the river water percolates into the soil and creates a strip with sufficient ground water to support a forest in an area where the predominant vegetation is grassland. These strips of forest may be likened to oases in a real desert, where supplies of subterranean water reach the surface and support a moisture-loving vegetation, or to the man-made oases in many deserts where irrigation provides the necessary water. Not every kind of forest plant can live in these strips along rivers or in an oasis, however. Those that do live there must be able to withstand exposure to hot dry air and strong winds.

There are even places where local climates differ so greatly from the general climatic pattern of the region that a special type of vegetation is developed. A notable example is the fog belt along the Pacific Ocean, extending for many miles north and south of San Francisco. In this region the general climate is dry, with rarely more than 20 inches of rain, and that falling almost entirely in the winter, while the summers are hot and almost cloudless. In the fog belt, however, the summers are cool and the rainfall is often 40 inches and probably up to 80 inches in especially favored spots. Naturally the vegetation is true forest instead of the sclerophyllous forest and chaparral of the dry hills. Neither the fog nor the forest ascends very high on the mountains. If a mountain is high enough to extend above the fog, its summit is covered with a different type of vegetation consisting mostly of sclerophyllous trees and shrubs. In other words, the top of the mountain is warmer than its base, which is just the opposite of one's normal expectation. There are few if any surprises more astonishing to the botanical tourist than the one he finds when he drives to the top of Mt. Tamalpais, just north of San Francisco, starting in the cool damp Redwood forest of Muir Woods and ending at the hot, dry, sunny summit covered with chaparral.

From Chapter 20, a reader might excusably derive the idea that all vegetation in the Temperate Zone (very little has been said about the tropics) is not only closely correlated with climate but also sharply divided into four different types separated by sharp boundaries. If that were true, a vegetational map would look something like a political map, where every boundary line is accurately defined by law or by treaty.

Not so with plants. Every kind of plant is a law unto itself, migrating wherever it can, contracting its range when it must. Contiguous provinces always overlap. The vegetation, and the kinds of plants which constitute the vegetation, change from place to place just about as gradually or as abruptly as the climate changes, and that is usually pretty slowly.

Possibly the most gradual change in climate is to be observed along a north-south line anywhere in the eastern third of the country from southern Canada to the Gulf of Mexico. Throughout this long distance the rainfall is always sufficient for good forest growth, and the only essential change is in the temperature. Even in that, there is comparatively little change in the summer temperature, but the northern winters are much longer and much colder, and the growing season is much shorter. This change is shown by a very gradual decrease in mean annual temperature stretched out over about 1,200 miles of distance. Along such a line, whether it is drawn from Duluth to Galveston, or Sault Ste Marie to Mobile, or Quebec to Jacksonville, there may be some abrupt changes in plant life, but these are explainable by differences in the soil or by the history of the vegetation or by both together.

Abrupt changes, on the other hand, are most conspicuous where mountains rise directly from the plain. Then in relatively few miles there is a decrease in temperature and usually also an increase in rainfall, and the two combine to produce correspondingly abrupt changes in the plant life. The most noteworthy example of this is along a line from the summit of Mt. Whitney, highest peak in the contiguous United States, to the bottom of Death Valley, which descends to more than 200 feet below sea level; from a mountain always cold at the summit, snow-covered most of the year, with essentially no vegetation, through coniferous forests of high rainfall, into and through desert with low rainfall and high summer temperature, and finally into Death Valley, with virtually no rainfall, with summer temperature frequently surpassing 120 degrees and rarely passing 130, and almost devoid of plant life. All of this change occurs in an airline distance of about 70 miles. Eastward from the high peaks of the Front Range of the Colorado Rockies, as Long's Peak and Pike's Peak, an even shorter distance makes almost as much change in the vegetation, which progresses from alpine treeless summits through coniferous forests and into grassland.

All this means that the observant traveler, as he rides or even flies across the country, may sometimes pass abruptly from one province into another, but is more likely to use many miles in making the transition. There are only two general rules about these transitions which can be stated now. The first has already been given: the more abrupt the change in climate, the more abrupt will also be the transition in vegetation. The second: If the dominant vegetation forms of the two adjoining provinces are the same, the transition is probably gradual; if they are different, as forest and grassland, the transition is usually more abrupt and is certainly more conspicuous.

In Chapter 20 a set of symbols was introduced by which the rainfall requirements of four types of vegetation might be indicated. Let us print them again, this time in the form of a square, and draw lines extending from each of them to the other three (with one exception, which the reader will note at once and which will be discussed farther on).

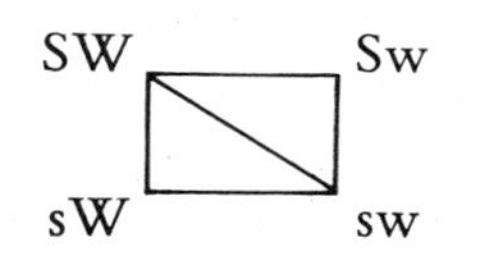

These lines denote the transition from one type of climate to another. If summer rain continues to be adequate while winter rain steadily decreases, we get the transition shown by the upper horizontal line, and the vegetation changes

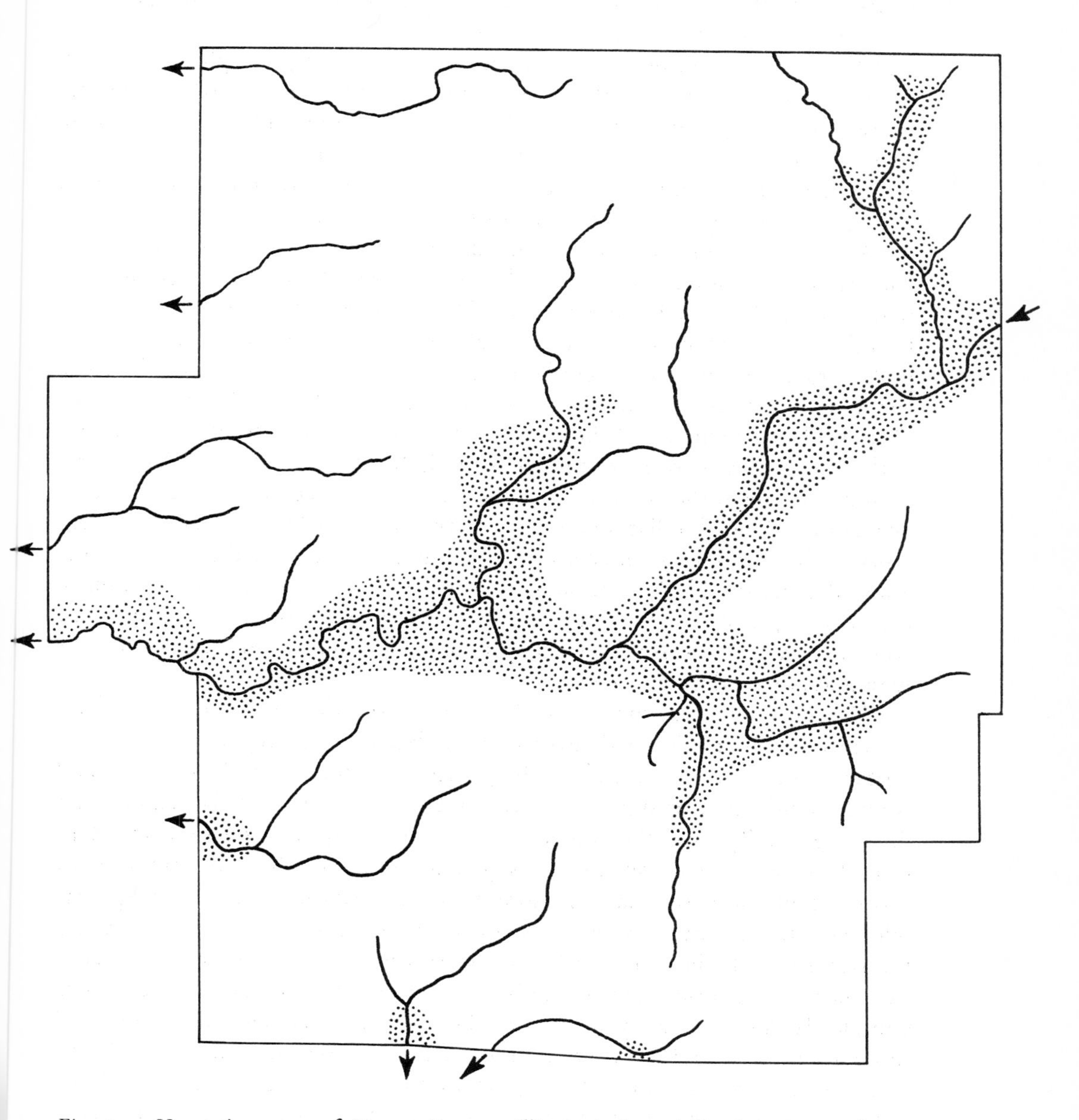

Fig. 21.2. Vegetation map of Macon County, Illinois, before civilization. Forested area stippled, grassland area blank.

from forest to grassland. If winter rainfall remains adequate while summer rain decreases, we have the transition shown in the left hand vertical line, in which the vegetation changes from ordinary forest to sclerophyllous forest. Similarly, we can see from the diagram the transitions to desert from normal forest, from grassland, and from sclerophyllous forest. Every one of these five transitions can be seen from major highways in the United States, and it is most interesting to note how the plant life changes during a drive of half a day or less. We shall consider some of them in detail.

Any person living in New England, New Jersey, or eastern New York and driving to Florida might reasonably use the New Jersey Turnpike through that state, U.S. 40 to Baltimore, and U.S. 301 thence to Florida. These last two highways run very close to the fall line, which marks the shore of the prehistoric ocean and the present boundary of the coastal plain. They therefore run along the transition line between the Deciduous Forest Province and the Coastal Plain Province and show more or less of the vegetation of each. But if the motorist will turn south-east from U.S. Highway 301 at Petersburg, Virgina, and take the road to Norfolk, he will at once enter a new type of vegetation. The familiar kinds of eastern trees quickly disappear and are replaced by various southeastern species, among which are the live oak, the longleaf pine, the black gum, and the bald cypress. Mistletoe is common in the trees; Spanish moss is soon observed hanging from the branches. The woods contain many shrubs and herbs completely unknown farther north or west and are tangled with vines.

The transition from the Eastern Deciduous Forest to the Prairies and Plains Province has been discussed several times already. At present, it is scarcely visible to a traveler because all of the prairie land and much of the forest land is under cultivation or supports houses and streets. Since grassland and forest do not generally mix, the contact was a very sharp one, where a few steps took the observer from one type of vegetation into the other. But this did not form a continuous or relatively straight line, separating forest from prairie once and for all, but a series of lines, deviously arranged and aggregating thousands of miles in total length. At the eastern end of the Prairies and Plains Province, the prairie occupied only isolated patches of varying size, these completely surrounded by forest. Farther west the patches were larger and more numerous, and in western Indiana and Illinois they were confluent. There the forest was reduced to strips along the streams. These strips could be called continuous by stretching the definition of the word, because those of two neighboring streams finally met and were united at the confluence of the rivers. In their upper courses, the forests of one stream were separated from those of the next by a broad or narrow strip of prairie, and these could often be followed around the source of the rivers where they became confluent with the next prairie. It will be remembered from Chapter 20 that all of this transitional area has a forest climate and that the prairie is merely a relic, a large and important one to be sure, of a former time when the climate was considerably drier. Never forget the importance of past vegetational history in explaining the present distribution of plants.

This transitional condition continued across Illinois and Iowa, with a steady decrease in the number of isolated bits of forest and in the width of the forest belts along the streams. In northwestern Iowa and not far beyond the Missouri River in Nebraska and Kansas, the forest could no longer grow on the uplands but was confined to the alluvial land near the river bed. This marks the climatic limit of the forest. Forests continue considerably farther west, but only in narrow and often interrupted strips close to the water, where the wet soil compensates for the lack of rain.

This state of affairs is hard to imagine today by a motorist as he rides west on such highways as U.S. 30 and 36 across Illinois and Iowa. He will see some wood lots still remaining, never very far from the rivers and creeks, and he can distinguish where the prairie used to be by the black color of the soil, if the land is not concealed by growing grain.

One has scarcely passed the last cottonwood tree along the Platte River before low hills break the monotony of the plains, where the altitude is already more than 3,000 feet. At the western end of Nebraska (not of Kansas) he may catch sight of a few outlying groves of ponderosa pine (*Pinus ponderosa*), the first tree of the Cordilleran Forest Province. The actual transition can be well seen on many Colorado roads in the vicinity of Boulder, Denver, Colorado Springs, or Pueblo. The margin of the forest is sharply defined. A few forest shrubs and herbs, but very few, are growing among the grasses, and several plains plants grow under the pines. A few miles of added distance westward brings a considerable rise in altitude, and with it a decrease in temperature and an increase in rainfall, and only these few miles are needed to complete the change from grassland into forest.

One of the best places to see the transition from forest to desert is on U.S. Highway 66, driving either east or west across Arizona. After entering the state, either from New Mexico or from California, the highway passes for miles through real desert. Then the altitude begins to increase as the road approaches the Arizona Divide, and in a few miles trees of juniper can be seen scattered among the desert plants. They become more numerous and taller, and scattered trees of ponderosa pine appear among them, while the desert plants disappear. The pines become larger and more numerous until they form a regular forest, and among them appear scattered trees of Douglas fir. These in turn gradually become more numerous, occupying more space at the expense of the pines, until they too form a forest. After several miles the road starts downgrade. Pines appear again and are soon replaced by juniper; desert plants appear among the junipers; the last junipers are passed, and the highway is once more in the desert.

This transition will be of especial interest to the plant geographer, because both directions can be seen in a single trip. It is a matter of common experience that the disappearance of a particular kind of plant is seldom noted, while the appearance of a new one is more conspicuous and more frequently observed. One going up the grade may not notice where he saw the last juniper, but coming down he will easily notice the first one.

Not all provinces are distinguished by the nature of their rainfall. In the eastern states there are three forest provinces, all with ample rainfall both winter and summer. Although the Coastal Plain Province averages warmer than the Eastern Deciduous Forest Province, the most important line separating them is a geological one, the contrast between old soils and new, which is, again, a matter of history. The Northern Conifer Province and the Eastern Deciduous Forest Province both have a great variety of soils and both are located largely or wholly on glaciated territory. The present difference between them is mostly a matter of temperature, but the reason for the existence of two different types of forest side by side is a historical one and goes back farther than the glacial period.

Beginning at its northern terminus at the extreme north end of Maine, U.S. Highway 1 passes through forests of fir, spruce, and pine, with bogs occupied by tamarack and black spruce. Aspen and paper birch are common almost everywhere. Farther south scattered groves of sugar maple appear, and beyond Portland (southwest) the last of the fir, spruce, and tamarack have vanished, although aspen, birch, and some white pine may be seen even into Massachusetts. In place of these northern trees, the forests are now composed of beech, sugar maple, red maple, ash, elm, and various kinds of oak.

The road southeast from the great bridge over the Straits of Mackinac in northern Michigan, leading to Detroit, passes at first through great stretches of land covered by aspen and birch, indicating where forests of white pine formerly stood. Some pines (white, Norway, jack) will still be seen; some fir and spruce, but not much, and in the bogs plenty of tamarack. Even this far north there are already scattered forests of sugar maple and beech. Before the road reaches Bay City most of these northern trees have disappeared, and from there to Detroit the land was formerly covered with deciduous forests which continued on with little change for some hundreds of miles to the south.

One line was missing from our little diagram showing vegetational transitions, the line from Sw to sW. If one will think about this for a moment, he will realize that such a transition is theoretically impossible. As the S decreases to s and the w increases to W, there must be a place where the two are equal. Since the resulting rainfall is equably distributed between summer and winter, the resulting vegetation should be either desert if both s and w are small enough, or forest if both are large enough, or a transition between forest and desert. Therefore, to get from a sclerophyllous forest to grassland, one must, at least theoretically, pass through another type of vegetation.

As elsewhere, the theory is not always realized in practice. An example may be seen by the traveler who rides up the highway through the middle of the Great Valley of California and then turns east into the Sierra Nevada. The valley floor, now largely under cultivation, was formerly occupied by grassland. As the land rises into the foothills of the Sierra, the grassland is still evident and passes directly into sclerophyllous forest. In the mild climate of the valley, growth of plants is possible at all times of the year. These grasses start to grow in winter, just as soon as the winter rains have provided enough water in the soil. Even in December the low hills are often turning green. The grasses bloom early, and die and turn brown during the long, hot, rainless summers. The summer is their dormant season, but it is determined by drought rather than by cold. It must be emphasized, however, that the rainfall in both types of vegetation comes during the winter, and that the grassland is an unusual type quite different from the grassland of the Great Plains. There is here no transition from Sw to sW, but only from a special grassland to sclerophyllous forest.

In summary: The various provinces and floristic groups are separated, not by narrow lines, but by transition zones of varying width. In these zones the transition may be shown in either of two ways, either by the actual mingling of species from the two floristic groups or by the mingling of colonies of the two types. In the former case, the plants of the floristic group which we are leaving change from very numerous to many to several to few to none, while those of the group which we are approaching change from none to few to several to many to very numerous. Or, in the latter case, the colonies of the vegetation which we are approaching change from none to few and scattered, to larger and often confluent, and finally to continuous.

It has been necessary to describe these transitions largely in general terms. Few printed descriptions are available which describe any of them in great detail. Mention has already been made in these pages of observations in northern Minnesota, where a transition zone 160 miles wide separates the fir forests of the north from the maple forests of the south. John T. Curtis, in his magnificent book, *The Vegetation of Wisconsin* (1959), has discussed a transition zone in considerable detail and has plotted on a map the northern boundary of many southern plants and the southern boundary of many northern ones. The most concentrated change in the kinds of plants is along a belt extending from northwest to southeast across the state, which may be as narrow as 10 to 15 miles. This is not a transition from the Eastern Deciduous Forest to the Northeastern Conifer Province, but rather between two divisions of the former. South of this zone the forest trees are largely oak and hickory; north of it maple, beech, and hemlock, with a copious admixture of pine.

Original land surveys, made in most parts of the Middle West before much of the land had been put under cultivation, often show the distribution of forest and prairie in great detail. The surveyors ran their lines north-south and east-west a mile apart. They noted where the margin of forest crossed these lines and sketched in by eye the interior part of each square mile, and that was not difficult on the level prairies of Illinois and Iowa. The accompanying map (Fig. 21.2) shows the distribution of forest and prairie, as recorded in these early surveys, for Macon County, Illinois, and demonstrates how the forest was located in belts along the principal streams. Today U.S. Highway 36 runs east and west across the center of this county, and from it an observer can demonstrate the accuracy of the old map. It would appear from this map alone that the prairie in the northern part of the county was completely separate from that of the southern part. However, if one traced this prairie to the northeast beyond the limits of the map, he could follow it to the headwaters of the river (the Sangamon), pass the last extremity of this strip of forest, turn back southwest along the left bank of the river, and soon reach the southern prairie shown on the map.

Finally, as a last illustration of transitions in plant life, we go to Puerto Rico and drive down the well-made and very scenic road from Cayey to Guayama. This road passes over the mountains at an altitude of over 2,000 feet, where the annual rainfall is 60 to 80 inches and the vegetation was originally luxuriant forest. It then descends, with steadily decreasing rainfall, to sea level, where the vegetation is almost a desert. The following statement is taken, with only slight modification, from H. A. Gleason and Melville T. Cook's *Plant Ecology of Puerto Rico* (1927).

The southern boundary of the mountain vegetation is broadly determined by a general moisture relationship in which rainfall, atmospheric humidity, insolation, and wind are doubtless concerned. Its actual location in detail is also regulated to a greater or lesser extent by local conditions of soil and topography. Each ravine which descends southward from the mountain tends to have a moister soil, as it collects water from runoff or seepage, and to offer less exposure to the drying effect of sun and wind. Each shoulder of the mountain, fully exposed on its convex southern side to sun and wind, tends to be more xerophytic. The ravines therefore carry a mesophytic vegetation to levels below the general average, while the shoulders are occupied by xerophytic vegetation to a height considerably greater. If the observer should descend the steep slopes of the mountains directly, he would pass once from the mesophytic vegetation of the mountain into the xerophytic pastures of the foothills, at a high elevation if on a shoulder, or at a low altitude if he were following a ravine, and would continue through the same xerophytic vegetation to sea level. The highways are built with a more gentle gradient and descend the mountain on a long incline that crosses several shoulders and ravines before reaching the bottom. They therefore cross from one vegetation into the other several times, entering the xerophytic type as they round the shoulders, and reentering the mesophytic thickets as they turn into and cross the ravines, until they eventually descend below the last outpost of the mountain forest.

The plants seen along this road are of course completely strange to residents of the United States, although some of them belong to familiar genera. One of the most conspicuous of the mountain mesophytes is the terciopelo (*Heterotrichum cymosum*), a tall shrub or low tree abundant at higher altitudes. Its broadly heart-shaped leaves are thickly beset, especially when young, with red-purple hairs which make the plant both conspicuous and attractive. The ucar (*Bucida buceras*) is the most abundant and conspicuous tree of the dry lowland scrubby forests.

The overlapping and interdigitating margins of our provinces pose a real problem to the map maker. We are used to seeing our maps with the states marked by hard and fast lines and possibly colored with different tints. That is misleading when provinces overlap by a hundred miles or more. We could of course draw any one province and bound it by a definite line, provided we could agree where this line should be placed. Suppose, for example, that we wish to map the Northern Conifer Province. In the longitude of Indiana and Illinois there is a belt about 400 miles wide where plants of this province share the ground with plants of the Eastern Deciduous Forest. In northern Indiana and Illinois there are scattered groves of white pine with their attendant northern plants growing beneath them, and peat bogs with the typical northern tamarack, pitcher plant, and leatherleaf. Shall we draw our line far enough south to include all these outlying colonies? There are fine forests of eastern sugar maple and beech as far north as Lake Superior. Shall we draw our boundary far enough north to include only the lands where the northern forests have a complete monopoly? Or shall we take a middle course and locate the boundary across the center of Wisconsin and Michigan, where the northern forests begin to form a conspicuous feature of the landscape? And as to the northern boundary of the Northern Conifer Province, we are so unfamiliar, through lack of exploration, with all the devious crooks and turns of the tree line across Labrador, Yukon, and the Northwest Territories that about the best we can do is draw a sweeping line across northern Canada. Then we must remember the principle of discontinuous distribution, as discussed in Chapter 4 and excellently illustrated by the map of Macon County in this chapter. The same condition mapped there extends over some 200 other counties in the Middle West. To get the facts would require long and patient delving into old surveys and old literature, and to show these facts would require a map on an impracticably large scale.

So, when one sees a vegetation map of the world or a continent in an atlas, and many atlases have such maps, he must remember that discontinuous distribution operates in every continent, that transition zones between different floristic groups are typical of all countries, and that those distinct lines on the vegetation maps would be more accurate if they were broadened into stripes. But whether they should be broadened on one side or the other or both, the unfortunate reader can only guess.

The authors can only hope and trust that the reader, after having patiently floundered through this chapter, will still remain convinced that floristic groups actually do exist. They are a fact. They do exist. One need only travel over the country to see them for himself. Then the reader can join with the student quoted many pages before and say with him, "It's all true!"

22/ The Vegetation of North America, North of Mexico

We have seen that climate is the principal factor determining the general nature of the vegetation of the various parts of the earth. The kind of soil also has some influence, but to a very large degree the soil type is also governed in the long run by climate, both directly and through the vegetation. The coastal plain of eastern and southeastern United States is the only large area in the country where the soil type has a strong effect on the vegetation independently of climate, although there are many small areas where the soil type is locally decisive. We have emphasized the importance of history in plant geography, but this affects the distribution of individual species more than the life form of the plants occurring in a given region. We thus return to climate as the over-riding control on natural vegetation over broad areas.

When the plant cover is removed or disturbed by some catastrophic force—fire, hurricane, glaciation, vulcanism, or man—the nature of the vegetation may be changed for a greater or lesser period of time. Typically the first plants to take over a newly available area are those which have excellent means of seed dispersal, even though they may not be able to compete well with other plants under more stable conditions. Sooner or later, in the absence of further disturbance, the first invaders are replaced by other species better adapted to prolonged competition, and there may be several different successive communities before the vegetation returns to the normal type for the particular climate. This climatically controlled vegetation type in any area is referred to as the regional climax. All other communities in the area tend to change toward the regional climax, in the absence of disturbance, although completion of the process may take a very long time under some conditions. Even the lakes eventually get filled in by the deposition of sediments and vegetative debris.

It is the climax communities with which the following discussion of the vegetation of the various provinces will be primarily concerned. Seral communities (i.e., those which are part of a series leading toward the climax) will be discussed mainly when they are common and slow to change.

It may be argued whether certain communities are seral or climax, and much depends on the definitions one starts with. Any individual swamp is transitory in terms of geologic time, but in terms of human time it may be permanent, and it may well outlast the climate which now prevails. A flood-plain along a stream will retain its character as long as the stream continues to act on it, and the plant community which inhabits it will continue to reflect the particular ecologic factors involved. The climate is, of course, an important factor influencing the vegetation of such special habitats, but here it clearly shares the control with special factors relating to those habitats. Our discussions will be concerned with such long-persistent special communities only when they are distinctive enough and common enough to catch the eye of the observant traveler.

We have recognized ten floristic provinces in the continental United States and Canada (shown in Fig. 15.1). As we have pointed out, these provinces mostly grade into each other, rather than being separated by clear-cut, narrow lines. The several provinces are discussed in the following sequence: 1, Arctic or Tundra Province; 2, Northern Conifer Province; 3, Eastern Deciduous Forest Province; 4, Coastal Plain Province; 5, West Indian Province; 6, Prairies and Plains or Grassland Province; 7, Cordilleran Forest Province; 8, Great Basin Province; 9, Californian or Chaparral Province; 10, Sonoran Province.

THE TUNDRA PROVINCE

The northernmost parts of North America and Eurasia, north of the boreal forest, belong to the Tundra Province. The most characteristic features of the arctic tundra are the absence of trees and the presence of permafrost (permanently frozen soil) at a depth of a few inches to a few feet. The depth to which roots penetrate is governed by the depth to which the ground thaws in the summer. Much of the ground surface is spongy and hummocky, because of the winter freezing and summer thawing. The freezing often results in frost boils, which may be as much as 30 feet wide and 3 feet high. In the moister areas, the ground often cracks into rough polygons 15 to 25 feet across (Fig. 22.1), the cracks being filled with deep wedges of ice which thaw at the surface in the summer.

The growing season in the tundra region is short, generally only 2 or 3 months, and there may be frost any day of the year. Even during the brief summer there is often much fog and mist. During clear weather the temperature may rise as high as on any summer day in New York, and the long or continuous days permit rapid growth.

Fig. 22.1. Patterned ground near Point Barrow, Alaska; aerial view. Note vehicle tracks in the lower picture. (Photos courtesy of W. C. Steere)

Fig. 22.2. White heather (*Cassiope Mertensiana*), a species characteristic of high altitudes in the mountains of western Canada and the United States. A closely related species, *Cassiope tetragona*, is circumboreal in the Arctic. (Photo by Gayle Pickwell, from National Audubon Society)

The precipitation in tundra regions is not high, but the potential evaporation is generally even less, and the runoff is usually slow. Much of the land is flat and poorly drained, with a sticky, typically blue-gray soil that is rich in organic matter and more or less saturated with water. Mountainous parts of the arctic pull down more snow in the winter than the lowlands and are likely to be continuously ice-covered. Even in summer only a narrow fringe of Greenland is ice-free, and the enthusiastic Viking who named it would have been a suitable head for a Chamber of Commerce.

In much of the arctic the vegetation is in very precarious balance with the environment, and seemingly minor disturbance can have major effects. Caribou often destroy the vegetation locally by grazing and trampling. On the other hand, the plant cover is apt to be more luxuriant on abandoned Eskimo village sites, because of the higher concentration of nitrogenous matter in the soil. Such a seemingly unimportant thing as a vehicle trail may establish new drainage channels and cause considerable erosion. A competent botanist observed in 1962 that "activities of white men during the past 15 years have done more to change the face of the arctic tundra than had been done by the Eskimo inhabitants during their entire history."

The tundra flora consists of relatively few species, but most of these are widespread and many are circumpolar, extending all the way from northern Scandinavia through Siberia and North America to Greenland. The paucity of the flora may not be evident in any one place, since there may be as many or even more species per acre than one would find in California, Iowa, or New York. The same things occur monotonously, however, over acre after acre for mile after mile, from Ungava to Point Barrow and on around the pole. A recent treatment of the flora of the Alaskan arctic slope includes only 419 species of seed plants, but well over half of these are widely distributed in the arctic.

Grasses, sedges, small flowering herbs, low shrubs, lichens, and mosses are the principal elements of tundra vegetation. Toward the south the plants cover most of the soil, and shrubs, especially low evergreen members of the heath family (Ericaceae), are not uncommon. Much of the land may have a brilliant blanket of flowers in July and August, and thickets of birch, willow, and alder may be taller than a man along the streams.

Northward the plant cover is progressively sparser. Mosses and lichens make up an increasing part of the community, not so much by an increase in their own numbers as by a decrease in the other plants. Displays of showy flowers are more restricted to south slopes or otherwise protected spots, and a willow thicket may be only an inch or so high, resembling a clover lawn. Grasses, sedges, and peat moss are common in the moister spots.

Perhaps the most characteristic single genus of the arctic tundra is *Cladonia* (Fig. 22.4), the "reindeer moss," actually a lichen. It is commonly a few inches tall and much branched, like a diminutive shrub. Among the true mosses, the hairy-cap moss (*Polytrichum*) is common in dry places and peat moss (*Sphagnum*) in wet ones. *Carex* and *Eriophorum* (cotton grass, Fig. 22.5) are common sedges, and *Poa* (bluegrass) is a common grass. Buttercups (*Ranunculus*), louseworts (*Pedicularis*), saxifrages (*Saxifraga*), Jacob's-ladder (*Polemonium*), daisies (*Erigeron*) and campions (*Silene*) are common wild flowers.

To the south, the arctic tundra gives way to the boreal forest in the lowlands and merges with the alpine tundra in the mountains. The timber line which marks the lower limits of the alpine tundra corresponds to the tree line of the far north. The northern tree line, however, unlike the often fairly sharp timber line of the mountains, commonly spreads over a broad belt, often a hundred miles wide, in which trees are progressively restricted to the most protected habitats. On our map the southern boundary of the Tundra Province is not placed at the absolute limit of trees, but more nearly where there is an equal division between low forest and tundra.

The vegetation and environment of the alpine tundra resemble those of the arctic tundra at first glance, but there are significant differences. Alpine tundra is not generally underlain by permafrost, and the topography is generally rougher, so that much of the land is better drained. The growing season is longer, but the days are shorter. Mosses and lichens are much less prominent parts of the vegetation, and the flower-garden aspect is often even more pronounced. Many species are especially adapted to the rock crevices and talus slopes which are so abundant in the high mountains.

Most of the common genera and even many of the species of flowering plants in the alpine tundra of the United States are the same as, or closely related to, those of the arctic, and a person familiar with the flora of one would also feel at home in the other. The differences become progressively greater as one travels southward, however. There is a good deal of alpine tundra in Colorado, but only a few of the species are the same as those of the far north. Many of the species, although not actually identical, are closely related to those of the arctic, but many others are alpine derivatives of plants from lower elevations, some of them belonging to genera which do not reach the arctic at all.

Fig. 22.3. Tundra near Point Barrow, Alaska. (Photo courtesy of W. C. Steere)

Fig. 22.4. Reindeer moss (*Cladonia*). The intricate, bushy branching is characteristic of the genus. (Photo by Helen G. Cruickshank, from National Audubon Society)

Fig. 22.5. Cotton grass (*Eriophorum*) meadow in interior Alaska. (Photo by Charles J. Ott, from National Audubon Society)

Fig. 22.6. Alpine tundra in Colorado. Some of the species growing in the foreground, such as *Geum Rossii*, are widespread in the arctic as well as at high altitudes in the western cordillera; others, such as *Trifolium nanum*, are Rocky Mountain alpine species which do not reach even as far north as the Canadian border. (U.S. Forest Service photo by H. E. Schwan)

The alpine tundra (Fig. 22.6) is restricted to progressively higher elevations as one travels southward in Canada and the United States. In the east, it barely reaches the higher mountains of New England and northern New York. In the western cordillera, it is well developed on the mountain tops as far south as central New Mexico. The actual elevation of timber line varies with the snowfall as well as with the latitude. Other things being equal, 1 mile of latitude in temperate regions equals about 4 feet of altitude, but in the dry mountains of central Idaho the timberline is near 10,000 feet, in contrast to less than 7,000 feet on Mt. Hood in the moister Cascade Mountains at the same latitude in Oregon.

In Mexico and Central and South America the flora above timber line is in general very different from the alpine tundra of high mountains in the United States. The paramos, as these areas above timber line in the tropics are called, cannot be included in the Tundra Province, and are outside our field of discussion.

THE NORTHERN CONIFER PROVINCE

The Northern Conifer Province forms a broad belt, up to 500 miles wide, south of the tundra in Canada and Alaska, dipping into the contiguous United States around Lake Superior and in the mountains of northern New England and New York. The general trend of the province is somewhat southeast-northwest rather than straight east-west, reflecting a similar trend of isotherms on climatic maps. The east coast of Canada and northern United States is cooled by the Labrador Current, whereas the comparable parts of the west coast are warmed by the Japan Current.

The Northern Conifer Province is, as its name implies, dominated by coniferous forest. The dominant, widespread plant community is often referred to simply as the boreal forest. The province extends across Siberia and into northern Europe, but it is less homogeneous than the Tundra Province. Not many of the common species of the boreal forest in Sweden are actually the same as those in Canada, but nearly all of the genera are.

Much of the boreal forest region is a land of low relief, with many lakes and slow streams and extensive boggy areas. The bogs are often more or less covered with sphagnum moss and are then called muskegs (Fig. 22.7). Away from the bogs, the soil is typically pale and sandy beneath a layer of slowly decaying litter and humus. Such soils, called podzols, are not very fertile. They are somewhat acid and are low in available mineral nutrients. Under the climatic conditions of the Northern Conifer Province the ferromagnesian

Fig. 22.7. Muskeg in southern Alaska. (U.S. Forest Service photo by Linn Forrest)

minerals leach out of the soil much more easily than does silica, and podzolic soils develop as a result. Only a small proportion of the land in the Northern Conifer Province is cultivated.

The climate of the Northern Conifer Province is cold and moist. The precipitation is not very high, commonly about 15 or 20 inches per year, but it is still more than the potential evaporation. The growing season is only 3 or 4 months long, and there may be frost at any time. Winters are long, cold, and snowy.

The number of kinds of plants which are adapted to such conditions is not large, and in North America only the Tundra Province has fewer species. A plant which can survive in one part of the province can probably also do so in another, however, and species after species ranges all the way from Quebec to Alaska. Many of the same plants occur also in Siberia, and some reach all the way to Scandinavia.

The trees of the boreal forest are small, seldom over 50 feet tall or with a trunk over 2 feet thick, but they may form a very dense cover. One mile an hour is fast—and exhausting—progress for a man making his way through such a forest, ducking under low branches, climbing over fallen trees, circling to avoid perilous muskeg, and searching for nonexistent landmarks to keep his bearings. In drier areas, on the other hand, the trees may be well spaced, with little or no undergrowth.

The most characteristic tree of the boreal forest in America is white spruce (*Picea glauca*, Fig. 22.8). East of the Rocky Mountains, it is commonly associated with balsam fir (*Abies balsamea*, Fig. 22.9), and one thinks immediately of the spruce-fir forest (Fig. 22.10) when the Northern Conifer Province is mentioned. Tamarack (*Larix laricina*) and black spruce (*Picea mariana*, Fig. 22.11) are also common in this province, especially in the wetter sites.

Drier places in the Northern Conifer Province often carry a forest composed mainly of jack pine (*Pinus Banksiana*, Fig. 22.12). Like the lodgepole pine of the Cordilleran Forest Province, jack pine is often a fire tree. The small cones tend to remain on the branches, unopened, for many years. They do not burn readily, but after a fire they open and release their seeds. The bare ground left by the fire is an excellent seed bed for the pines, and a mixed forest with only scattered jack pines may be replaced by a nearly pure forest of jack pine after a fire. Under more stable conditions, in the absence of fire, jack pine tends to be replaced by spruce and fir in most places. In some of the most barren soils, however, it may persist and perpetuate itself indefinitely. Jack pine is not wholly confined to the Northern Conifer Province. It occurs also in the

Fig. 22.8. Twigs and cones of white spruce. Like most spruces, this species has sharply pointed, four-angled needles. (U.S. Forest Service photo by W. D. Brush)

Fig. 22.9. Balsam fir trees in northern Minnesota. The narrow, spire-like crown is characteristic of this and several other kinds of fir. (U.S. Forest Service photo by R. K. LeBarron)

Fig. 22.10. Spruce-fir forest in Quebec. The density of the forest is typical. (Photo courtesy of Pierre Dansereau)

Fig. 22.11. Thrifty, eighty-year-old stand of black spruce in Minnesota. Note the density of shade within the forest. (U.S. Forest Service photo by R. K. LeBarron)

Fig. 22.12. Jack pines in northern Minnesota (foreground). Note that both the pine and the birch (background) are being replaced by a stand of spruce-fir. (U.S. Forest Service photo by R. K. LeBarron)

northernmost parts of the Eastern Deciduous Forest Province, likewise mainly as a fire tree and mainly in sterile, sandy soils.

Three species of broad-leaved trees also occur in some abundance in the Northern Conifer Province of North America. These are the paper birch (*Betula papyrifera*, Fig. 22.13), the quaking aspen (*Populus tremuloides*, Fig. 22.14), and the balsam poplar (*Populus balsamifera*). They are especially common in burned or otherwise disturbed areas, but they often occur in moister sites than jack pine. Like jack pine, they generally tend to give way in time to spruce and fir, in the absence of disturbance. Their small, wind-distributed seeds give them a considerable advantage over the heavier-seeded spruce and fir in reaching newly opened habitats. Aspen also spreads extensively by the roots. Push over a dead aspen and you will usually find one large root running parallel to the ground in one direction, and another in the opposite direction. If you follow these a few feet you will find that each connects to another tree. Aspen seedlings are in fact rarely seen. Doubtless every aspen grove developed originally from one or several seedlings, but this is an inference rather than an observation. After a fire new trees come up again from the roots. Many aspen groves are doubtless centuries old, even though the individual trees generally live only a few decades.

Showy flowers are not numerous in the dense forests of the Northern Conifer Province, but some of those which do occur are very attractive, and the more so because there is so little greenery on the ground in general. Several members of the orchid family, such as *Calypso bulbosa* (Fig. 22.15), are gems of beauty. The Heath family (*Ericaceae*) is also well represented in the Northern Conifer Province. Several species of wintergreen (*Pyrola*), in particular, are fairly common. Members of both the Heath and Orchid families are commonly mycorhizal; i.e., their roots harbor fungi which also extend out into the soil and absorb dead organic matter. The nature of this partnership is not fully understood, but some members of each of these two families have become so dependent on their mycorhizal fungus that they have lost the ability to make their own food and do not have the green coloring matter (chlorophyll) found in most plants. Many other kinds of plants also have mycorhizal fungi, but in most instances the relationship is not so necessary as it is for these two groups. To grow an orchid or a heath, you must first provide the proper conditions for its mycorhizal fungus (Fig. 22.16). Bacterial decay proceeds slowly at the low temperatures characteristic of the boreal forest, and the fungi, including mycorhizal fungi, come into their own. The Orchid and Heath families are both well represented in other parts of the world as well as in the boreal forest,

Fig. 22.13. Paper Birch in northern Minnesota. Note that the young trees which will replace the birch are conifers, probably balsam fir. (U.S. Forest Service photo by R. K. LeBarron)

Fig. 22.14. Quaking aspen, with an understory of balsam fir, in northern Minnesota. (U.S. Forest Service photo by Harold O. Batzer)

Fig. 22.15. Calypso bulbosa. This characteristic orchid of cool coniferous woods occurs in Eurasia as well as in America. The scale can be judged by the mosquito at the right of the flower. (Photo by R. D. Muir, from National Audubon Society)

and the orchids in fact reach their greatest development in the tropics, but so many other familiar families are so poorly represented in the boreal forest that these two groups catch one's attention.

Another common flower in the boreal forest is the dwarf cornel (*Cornus canadensis*, Fig. 22.17), a colonial herb less than a foot tall, with flowers much like those of the well-known flowering dogwood (*Cornus florida*) of eastern United States. Various members of the Pink family (Caryophyllaceae), Buttercup family (Ranunculaceae), Saxifrage family (Saxifragaceae), and Rose family (Rosaceae), among others, are also frequently found in the boreal forest.

There is a broad transition zone between the Northern Conifer Province and the Tundra Province. As one goes northward in this *land of little sticks*, the trees are progressively smaller and more confined to protected sites. There is no sharp arctic tree line comparable to the timber line of mountains in the Temperate Zone.

On its southern margin the Northern Conifer Province adjoins three other provinces in different parts of the continent. None of the boundaries is sharp. The northern part of the Cordilleran Forest Province is scarcely more than a montane southern extension of the boreal forest; although the species are different, spruce and fir are still the dominant trees. Farther south, however, the Cordilleran Forest Province is much more complex, containing many elements that have little or nothing to do with the boreal forest. The spruce-fir forests extend nearly as far south as the Cordilleran Forest Province does, but they are progressively more confined to the upper elevations. As one climbs a mountain anywhere in the Rocky Mountain region, the uppermost zone of forest, just below timberline, is likely to be dominated by species of spruce and fir closely related to the white spruce and balsam fir of the north.

A similar but more attenuated arm of the boreal forest extends southward at increasing elevations in the Appalachian Mountain region as far as the Great Smoky Mountains of North Carolina. Here, as in the western mountains, the dominant trees are still spruce and fir, but the species are different from those of the boreal forest. White spruce gives way to red spruce (*Picea rubens*, Fig. 22.18), and balsam fir to Fraser fir (*Abies Fraseri*). Red Spruce also extends northward into the true boreal forest in southeastern Canada, but the Fraser fir is confined to the southern Appalachians. Some of the herbs and shrubs of the boreal forest also extend southward in the Appalachians with the red spruce and Fraser fir, but many of them do not. A good many of the species found there actually belong with the surrounding deciduous forest (Fig. 22.19). This southeastern arm of the spruce-fir forest is so narrow, interrupted, and

Fig. 22.16. Indian pipe (*Monotropa uniflora*), a colorless member of the Heath family which is widespread in cool coniferous forests. The plant is dependent on its mycorhizal fungus for food. (Photo by W. H. Hodge)

Fig. 22.17. Dwarf cornel (*Cornus canadensis*) in northern Michigan. This species is widespread in the cooler parts of North America and also extends into eastern Asia. (Photo courtesy of J. Arthur Herrick)

Fig. 22.18. Interior view of one of the few remaining stands of virgin red spruce in the southern Appalachians; near Durbin, West Virginia. (U.S. Forest Service photo in 1939)

Fig. 22.19. Rhododendron catawbiense, a characteristic species of the southern Appalachians, against a background of spruce and fir on Roan Mountain, North Carolina. (U.S. Forest Service photo by Daniel O. Todd)

floristically attenuated that for mapping purposes it is here included in the Eastern Deciduous Forest Province.

THE EASTERN DECIDUOUS FOREST PROVINCE

Most of the eastern part of the United States, exclusive of the southeastern coastal plain, belongs to the Eastern Deciduous Forest Province. This province covers a larger part of our country than any other single province. It extends also into southern Canada, particularly in southern Ontario and along the upper part of the St. Lawrence River. Nova Scotia and New Brunswick have a complex mosaic and partial blend of deciduous forest and northern coniferous forest elements, but on a broad-scale map without enclaves these political provinces are best assigned to the Northern Conifer Province, except for a small part of western New Brunswick which is continuous with the main deciduous forest of the United States.

The Eastern Deciduous Forest Province is limited on the north by progressively lower mean temperatures, especially as reflected in a shorter frost-free season; on the west by increasing aridity; and on the south and southeast by a sharp change in soil type at the geologic boundary of the coastal plain, as well as to some extent by increasing temperature.

Most soils of the Eastern Deciduous Forest Province are considered to be podzolic in the broad sense, but they diverge from typical podzol in the direction of laterite, the common soil of moist tropical regions. True laterites, which are hardly found at all in the United States, are characterized by leaching and loss of siliceous minerals and retention of iron and aluminum as oxides, so that the soil is a red or reddish clay of peculiar texture, which absorbs water easily and does not erode so readily as clay soils of temperate regions. Organic matter decays rapidly in tropical regions, and laterites contain but little humus. They are also low in available mineral nutrients, and they lose their fertility rapidly under cultivation.

The podzolic soils of eastern United States are more fertile than either typical podzol or typical laterite. The organic matter decays rapidly enough to provide a reasonable amount of humus, which is not so quickly destroyed as in tropical regions. Available minerals are more abundant than in either typical podzols or true laterites. Within these general limits the soils vary somewhat according to the latitude, the more northern ones approaching true podzols and the more southern ones approaching laterites. The climatic control is of course tempered by the nature of the parent rock. Soils derived from sandstone, for example, are different from those in the same region

Fig. 22.20. Spring beauty (*Claytonia virginica*), one of the common early spring flowers of the deciduous forest of eastern United States. (Photo by W. H. Hodge)

Fig. 22.21. Ginseng (*Panax quinquefolium*), a characteristic woodland herb in the Eastern Deciduous Forest Province, which has now become very rare because of its use in folk medicine. (U.S. Forest Service photo by Lee Prater)

Fig. 22.22. White snakeroot (*Eupatorium rugosum*), a common late summer flower in the deciduous forest. (New York Botanical Garden photo)

Fig. 22.23. Two common species of ragweed. Left, *Ambrosia trifida;* right, *Ambrosia artemisiifolia.* (New York Botanical Garden photos)

derived from limestone, and the difference is reflected to some extent in the flora. The influence of climate gradually reduces the difference, but especially in temperate regions one must expect to find continuing differences between soils of different origin.

The deciduous forest once covered hundreds of thousands of square miles in the eastern half of the present United States. Many of the trees were tall and stately, well over 100 feet high. Most of this virgin forest has now been destroyed. Farms, cities, roads, and suburban homes cover much of the land. Elsewhere the forest is in various stages of recovery after having been cut over one or more times. In some areas the plant cover has returned to something near its original state, except that the trees are not yet so large. In New England and New York a considerable amount of land that was once farmed has reverted to forest after being abandoned, and one finds old stone walls in the midst of dense forest.

The trees of the deciduous forest form an almost complete crown cover, and as a result very little direct sunlight reaches the ground in summer. There is generally a well-developed understory of shrubs and herbs, but walking through the forest is easy and pleasant. A number of low herbs, such as spring beauty (*Claytonia*, Fig. 22.20), anemone, hepatica (Fig. 14.11), trillium (Fig. 14.13), and violets bloom in early spring, before the trees have leafed out, and complete most or all of their growth within a few weeks thereafter. A succession of other plants bloom throughout the season, but these are not usually abundant and showy enough to produce a flower-garden effect.

From midsummer onward the most abundant and conspicuous flowers in the Eastern Deciduous Forest Province (e.g., Fig. 22.22) are members of the aster family (Compositae). Asters (Fig. 2.7) and goldenrods (*Solidago*, Fig. 11.1), in particular, are represented by numerous species as well as numerous individuals. More than a century ago, a German phytogeographer called our deciduous forest country and the adjacent coastal plain the realm of the asters and goldenrods.

Individual species of asters and goldenrods do not generally bloom throughout the whole season, but there is an overlapping succession of different species. *Solidago juncea*, which begins to bloom in July, is one of the earliest goldenrods throughout much of the province. *Solidago speciosa*, one of the latest species, is in full flower in October in New York.

Ragweeds (*Ambrosia*, Fig. 22.23) also belong to the aster family and bloom in late summer and early fall. Unlike so many other members of the Compositae, ragweeds have small and inconspicuous flowers. They produce great

Fig. 22.24. Cove forest (mixed mesophytic forest) in the mountains of North Carolina. The large trees in the foreground are, left to right, sugar maple, hickory, and buckeye. (U.S. Forest Service photo by Lee Prater)

Fig. 22.25. Large sugar maple tree, with a trunk about 4 feet thick, in the mountains of North Carolina. (U.S. Forest Service photo by Dan O. Todd)

Fig. 22.26. Birch-maple-basswood community in Wisconsin. The birch may be expected to diminish and disappear in the second-growth forest as time goes on. (U.S. Forest Service photo by Lee Prater)

Fig. 22.27. Oak-hickory forest in Virginia. This is a young stand, and the trees will become much larger. (U.S. Forest Service photo by E. S. Shipp)

quantities of pollen which is discharged into the air and is the principal cause of hay fever in most of the United States. Goldenrods, which bloom at the same time and are much more showy, have traditionally gotten the blame which properly falls on the ragweeds. The goldenrods are not as completely innocent as some of the recent reports would have us believe, especially inasmuch as some of the species do liberate some pollen into the air, but in general they are insect-pollinated instead of wind-pollinated, and they are only of very minor importance as compared to ragweeds.

Asters, goldenrods, and other composites are doubtless much more abundant now than they were in pre-Columbian times when the vegetation was less disturbed. Most of the species do not grow well if at all in dense forests; they are plants of open, sunny places instead. Even so, there could not be so many separate and distinct species in the region if they had not had satisfactory places to grow for thousands of years in the past.

The deciduous forest attains its most favorable development in the moist valleys and slopes of the southern Appalachian Mountains. More than two dozen kinds of trees are abundant and self-perpetuating under natural conditions in this mixed mesophytic forest (Fig. 22.24). Among these are species of basswood (*Tilia*), beech (*Fagus*), buckeye (*Aesculus*), hickory (*Carya*), magnolia, maple (*Acer*), oak (*Quercus*), and tulip tree (*Liriodendron*).

The number of common dominant trees in relatively undisturbed forest thins out in all directions from the southern Appalachian center. To the north, beech (Fig. 9.1) and sugar maple (Fig. 22.25) are increasingly abundant on the better soils, with oak and hickory often occupying the drier or more exposed sites. To the far northwest, in Wisconsin and Minnesota, the beech-maple community gives way to the maple-basswood community (Fig. 22.26) as the beech drops out and is replaced by basswood. The oak-hickory community (Fig. 22.27), confined to the poorer sites in the beech-maple and maple-basswood regions, is the most abundant type in the more southern part of the deciduous forest (outside of the Appalachian region). Toward the western edge of the oak-hickory forest, as one approaches the grassland, the trees become smaller and often more widely spaced (Fig. 22.28), and toward the southwest there is some savanna vegetation, with the trees widely scattered among the grasses. Farther north the forest-prairie transition zone generally shows a mosaic of the two vegetation types, as pointed out in an earlier chapter, rather than a blend.

East of the mixed mesophytic forest there was once a long fringe of oak-chestnut forest, extending from Georgia to southern New England. The

Fig. 22.28. Oak-hickory forest in Missouri. Note the relative openness of the forest. (U.S. Forest Service photo)

Fig. 22.29. Beech-maple-hemlock community in the mountains of Pennsylvania. The prominent trees in the foreground are hemlock (with rough bark) and beech (with smooth bark). (U.S. Forest Service photo)

Fig. 22.30. Mature stand of white pine in Minnesota. (U.S. Forest Service photo by A. Gaskill)

destruction of the chestnut since 1900 by the chestnut blight has changed much of this to an oak or oak-hickory community.

In much of New England, hemlock (Fig. 16.6) grows intermingled with broad-leaved deciduous trees, especially beech and sugar maple. This beech-maple-hemlock community (Fig. 22.29) extends west to the Great Lakes region, just south of the coniferous forest, and south in the Appalachian mountains to northern Georgia.

In pioneer days there were large stands of white pine (Fig. 22.30) in the same general area as the beech-maple-hemlock community. The pine forests developed mainly after fire or other disturbance, and each stand would tend to be replaced eventually by hemlock-hardwood forest. Most of the once extensive stands of white pine have long since fallen to the lumberman's axe. White pine also occurs in the more southern parts of the Northern Conifer Province in southeastern Canada and northeastern United States, as a fire-tree which eventually gives way to spruce and fir.

Other pines are found here and there in the deciduous forest region, especially after fire. Where the soil is poor and sandy they may persist more or less indefinitely, as jack pine does in northern Wisconsin and some other places.

Under some conditions other conifers may form a conspicuous or even dominant part of the vegetation in the Eastern Deciduous Forest Province. The well-known cedar glades of central Tennessee, characterized by eastern red cedar (*Juniperus virginiana*), occur in thin dry soil over flat-lying limestone rocks. The northern white cedar (*Thuja occidentalis*) is a common component of the dense, tangled forests of boggy places (Fig. 22.31), especially in the more northern parts of the province. In general, however, it is the deciduous forests which dominate the scene and give character to the landscape in the whole province, even under the disturbed conditions of the present time.

Bottomlands along streams are usually forested under natural conditions, but the trees are different from those of the upland forests. The eastern cottonwood (*Populus deltoides*) and the silver maple (*Acer saccharinum*), both of which can withstand prolonged flooding, are among the most common trees of flood plains (Fig. 22.32) in the Eastern Deciduous Forest Province. Some others which are often seen are American elm, black willow (*Salix nigra*), sycamore, sweet gum (*Liquidambar styraciflua*), and river birch (*Betula nigra*). The latter two species are especially common in the more southern part of the province.

We have mentioned the southern Appalachian region as having the most complex forest community in the Eastern Deciduous Forest Province. It also

Fig. 22.31. Northern white cedar (*Thuja occidentalis*) swamp in winter in Michigan. (U.S. Forest Service photo by G. B. Adams)

has a rich flora of shrubs and herbs. One reason for this is its geologic complexity. The Blue Ridge, including the Great Smoky Mountains, is composed mainly of ancient (pre-Cambrian) crystalline rocks, especially granite. Just to the west of the Blue Ridge is a strip of more recent (Paleozoic) sedimentary rocks of various sorts that are folded into long narrow ridges and intervening valleys. The westernmost part of the southern Appalachian region contains many uplifted plateaus with flat tops and steep sides. Much of the rock is flat-lying limestone. Lookout Mountain, on the borders of Tennessee, Georgia, and Alabama, is a typical part of this area. The flat top is now partly farmed, but the steep sides were a real barrier to transportation before the days of modern highways. The northwestern corner of Georgia (Dade County) is cut off from the rest of the state by Lookout Mountain, and during the Civil War this county seceded from Georgia, without any effective counteraction by that state. The varied topography and substrates of the southern Appalachian region thus provide a wide variety of habitats, and the observant traveler will see significant changes in the flora as he moves from one geologic province to another.

Another important factor in the relative richness of the southern Appalachian flora is that this area has been continuously available for habitation by land plants since late Paleozoic times, probably longer than the angiosperms themselves have existed. Northward, the land has been repeatedly glaciated during the past million years, and the most recent glaciation reached its height only about 11,000 years ago. The terminal moraine runs the length of Long Island in New York. Farther west the Ohio and Missouri Rivers mark the approximate southern limits of glaciation. To the south, east, and west of the Appalachian Mountains lies the coastal plain, which was under water at intervals during much of the Mesozoic and early Cenozoic geologic areas.

The Ozark Mountain region in southwestern Missouri and northwestern Arkansas is another floristic refugium, comparable in some respects to the southern Appalachians. The Ozarks are drier than the southern Appalachians, and this difference is reflected in the flora. Even so, there are a number of species-pairs in which one member centers in the southern Appalachians and the other in the Ozarks.

Revegetation of the glaciated part of the Eastern Deciduous Forest Province has proceeded mainly from the southern Appalachian and Ozarkian refugia, but not all of the species in these two areas have been able to migrate into this newly available region. Smaller parts of the present flora of the glaciated region have been derived from the coastal plain, the grasslands, and even the Rocky Mountains.

Fig. 22.32. Flood plain community of cottonwood and silver maple in the Des Moines River Valley, Iowa. (U.S. Forest Service photo by A. L. McComb)

Fig. 22.33. Virgin stand of longleaf pine in eastern Texas, with the characteristic carpet of grass under the trees. (U.S. Forest Service photo by Daniel O. Todd)

THE COASTAL PLAIN PROVINCE

The Coastal Plain Province occupies the geologic coastal plain of the Atlantic and Gulf Coast states, from New Jersey to Florida and the border of Texas, except for the southern tip of Florida. Like the geologic coastal plain, it also extends up the Mississippi River to the southern tip of Illinois. A progressively attenuated element of the coastal plain flora is found on southern Long Island (south of the glacial moraine), Cape Cod, and even the southern tip of Nova Scotia. Some coastal plain species occur in sandy soil around the southern end of Lake Michigan, perhaps reflecting a migration during a warmer time a few thousand years ago (the postglacial xerothermic period). Some other species once thought to be confined to the coastal plain are now known to occur locally in the southern Appalachians as well, and this latter area may indeed be their original home.

The coastal plain has in general the sharpest boundaries of any of our floristic provinces, because a clear geologic line coincides with a difference in vegetation which is governed partly by soil type. It would be much more difficult to draw the boundaries solely on the basis of the vegetation, both because of the progressive depauperization of the coastal plain flora toward the north and because many of the characteristic species of the Eastern Deciduous Forest Province also occur on the coastal plain.

The coastal plain is characterized by the presence of extensive pine forests. Northward, as in New Jersey, pitch pine (*Pinus rigida*, Fig. 13.3) is the dominant species; farther south loblolly pine (*Pinus taeda*), slash pine (*Pinus caribaea*) and longleaf pine (*Pinus australis*, Fig. 22.33) are more abundant. These are mostly rather small or medium-sized trees, although longleaf pine reaches a considerable size, up to 100 or even 130 feet.

The continued dominance of pines in this large area depends on repeated fires which have occurred for ages past and continue to occur. When the pine forest is dense it may be burned to the ground, but some of the species sprout again from just below the surface. Furthermore, fire hastens the opening of cones which have remained closed on the tree for years, and the newly released seeds germinate well in the bare mineral soil. Where the trees are more scattered, the ground is often covered by wiregrass (*Aristida*) and other low herbs which become tinder-dry in late summer and fall. Such areas may be burned every year without significant damage to the established trees.

Much of the coastal plain has a rather sterile, sandy soil which is low in mineral nutrients and does not hold water very well. In this soil the pines are at less of a competitive disadvantage vis-a-vis the broad-leaved trees than they are in the better soil farther inland, and the influence of fire becomes decisive. Whether the whole forest burns down or whether the undergrowth is merely burned off, the newly burned land is more suitable for pine seeds than for those of most broad-leaved trees, which generally prefer a soil with more humus and often need shade when young. The very abundance of pines insures that most of the tree seeds reaching the soil will be those of pines rather than hardwoods. These conditions reinforce each other, to the benefit of the pines.

In better soils farther inland, pines still occur in inhospitable sites and as fire trees, but fire is not enough to hold the balance permanently in their favor over large areas.

The interaction between fire, soil, and vegetation is well-illustrated by the inner pine barrens of New Jersey, which can be observed easily along state highway 72, a few miles southeast of its junction with highway 70. Here the soil is even more sandy and sterile than usual, and both pines and oaks appear to be mature when scarcely more than waist high. A little closer examination shows that these dwarf trees are really stump sprouts which have come up after fire. The base of the tree, just below the ground line, may be a foot or more thick and show traces of several previous stump sprouts which were burned back to the ground when only a few inches thick.

The inner pine barrens have obviously been subject to frequent fires for hundreds of years, but so has most of the rest of the coastal plain, and some additional explanation has been sought for the nature of the vegetation here. Repeated careful studies, however, seem to have eliminated all the logical possibilities not resulting more or less directly from fire. Geologically the inner pine barrens appear to be no different than the outer barrens, and all the major soil types from the outer barrens (reflecting details of geological history) are represented also in the inner barrens. The inner barrens are, however, slightly elevated, with no significant streams, swamps, or other natural firebreaks. Fire has probably swept over the inner barrens more frequently than over the other pinelands nearby, burning out the humus and keeping the soil relatively sterile. As a result, growth here is slower than elsewhere, and the crown of the trees does not get more than a few feet above the ground before another fire comes along. Since the main mass of easily combustible material is close to the ground, nearly everything is destroyed by the fire. Elsewhere, where growth is faster, the larger trees usually survive the frequent ground fires. The area where the fires keep the "forest" continuously stunted is essentially an area with a very high natural fire hazard, and the frequency of the fires has compounded their effect.

Fig. 22.34. Dwarf forest of pine and oak in the inner pine barrens of New Jersey, in April, 1963, before the oaks have leafed out for the year. This spot was last burned in 1947.

Fig. 22.35. Lowland forest, dominated by live oak, on Jekyll Island, off the coast of Georgia. Spanish moss hanging from the trees. (U.S. Forest Service photo by Daniel O. Todd)

Fig. 22.36. Loblolly pine (*Pinus taeda*), with an understory of hardwoods, in North Carolina. The pines are 80 to 90 years old. (U.S. Forest Service photo by Bluford W. Muir)

Fig. 22.37. Scrub oak woodland on sand hills in South Carolina. (U.S. Forest Service photo by Robert D. Shipman)

Fig. 22.38. Bald cypress pond in Florida, during an unusually dry season when the pond has dried up. This is probably *Taxodium ascendens*, often called pond cypress, instead of the more common and widespread, closely related species *Taxodium distichum*. (U.S. Forest Service photo by C. R. Lockard)

Fig. 22.39. Swamp with gum (*Nyssa*) trees in the foreground, near Disputanta, Virginia. (U.S. Forest Service photo by Bluford W. Muir)

Fig. 22.40. Pocosin in North Carolina, with a few stunted pond pines (*Pinus serotina*). If this site were slightly less wet, the pines would probably grow larger and be so abundant as to dominate the community. (U.S. Forest Service photo by Daniel O. Todd)

In the last several decades serious efforts have been made to prevent burning of the pine barrens, with the result that oaks are becoming more abundant in relation to the pines, and in the inner barrens it is now (1963) becoming difficult to find areas where the forest is not at least head high (Fig. 22.34) instead of merely waist high. The ultimate height of the trees in the inner barrens, if protected from fire, is uncertain.

Temperature also plays a significant part in molding the coastal plain flora. The more southern part of the province has a good many more species than the northern part. Many characteristic coastal plain species do not extend north of North Carolina, and the number diminishes progressively northward. The outer pine barrens of New Jersey, dominated by medium-sized or rather small pines and oaks, are perfectly good coastal plain vegetation, but they do not have nearly so many species as areas of similar appearance farther south.

When protected from fire, the pine forests of the coastal plain eventually give way, as we have noted, to hardwood forests. This replacement can be observed in progress at many places, and foresters have had considerable difficulty in growing succesive crops of pine on the same site.

In fairly moist, lowland but not swampy sites, the characteristic hardwoods which replace pines are beech, sweet bay (*Magnolia virginiana*, an evergreen species), and several species of live oak, notably *Quercus laurifolia* and *Quercus virginiana* (Fig. 22.35). Although sweet bay does not reproduce well under itself, it will come up under or after other trees of this community, as well as under pines; other trees, notably beech, will come up under the sweet bay. In addition to encroaching on the pines now, these several species dominate many small areas which have escaped fire in the recent past.

The more upland, better drained sites on the coastal plain were very largely pineland (where not cultivated) only a few decades ago, although there were oaks and other hardwoods in some places. More recently, since more attention has been given to fire control, these pine forests are beginning to be replaced by hardwoods (Fig. 22.36). The same hardwood species which take over the lowlands, plus several others which require a better drained soil, are the characteristic invaders. Especially common among these latter are two species of hickory (*Carya tomentosa* and *Carya glabra*), several deciduous species of oak (notably *Quercus alba*, *Quercus nigra*, and *Quercus falcata*), and holly (*Ilex opaca*). This mixed hardwood forest of upland sites on the coastal plain has more than a dozen dominant species, and compares in richness to the cove forests of the southern Appalachians. None of the species now occurs in all of the individual hardwood stands, but any of them may be expected almost

Fig. 22.41. Southern white cedar (*Chamaecyparis thyoides*) swamp in the pine barrens of New Jersey. The marsh in the foreground is dominated by a species of wool-grass (*Scirpus Longii*), a member of the Sedge family.

Fig. 22.42. Palmetto (*Sabal Palmetto*) grove in Florida. (U.S. Forest Service photo by C. D. Mell)

anywhere. Dry sand hills on the coastal plain often support a thicket of various species of scrub oaks (Fig. 22.37).

On the outer coastal plain the water table is seldom very far below the surface, and many areas are periodically or permanently flooded. Marshes with permanently standing water are usually dominated by coarse grasses and rushes. Places where water stands most but not all of the year commonly support forests of bald cypress (*Taxodium distichum*, Fig. 22.38). The irregular, knobby "knees" which arise from cypress roots and extend above the water have sometimes been thought to be aerating organs, but their anatomy does not support this view; no special function can certainly be assigned to them. In areas flooded for a shorter period, broad-leaved trees such as gum (*Nyssa*, Fig. 22.39) and ash (*Fraxinus*) are mixed with the cypress or replace it entirely. These hardwood swamp-forests, in turn, pass into the permanently wet but seldom flooded bogs, called pocosins (Fig. 22.40), which are typically dominated by evergreen shrubs and small trees, usually with a ground cover of Sphagnum. Burned-over pocosins often grow up to a long-persistent forest of southern white cedar (*Chamaecyparis thyoides*), which is the ecological equivalent, on the coastal plain, of the northern white cedar in the Eastern Deciduous Forest Province.

One of the characteristic species of low-lying but not really swampy sites on the outer part of the coastal plain is the dwarf palmetto, (*Sabal minor*, Fig. 17.8). This species, one of the few palms native to the United States, is more resistant to frost than most palms, but low temperature still limits its distribution, and it extends northward only to the coast of North Carolina. Unlike other palms, the dwarf palmetto is not a tree. Its stem creeps through the soil, and the stalks of the broad, fan-shaped leaves appear to arise from the surface of the ground. A related species, the cabbage palmetto (*Sabal Palmetto*, Fig. 22.42), is a tree sometimes as much as 50 feet tall. It is very common in Florida, where it grows in wet or dry places, in all kinds of soil, and in fresh, sulfur, and salt waters. Like the dwarf palmetto, it is limited in distribution by low temperature, and it barely reaches the coast of North Carolina.

Another characteristic coastal plain species whose northern limit appears to be determined by low temperatures is the Spanish moss (*Tillandsia usneoides*), which has been discussed in an earlier chapter. From southern Virginia southward it drapes over the limbs of various sorts of trees, sometimes even on telephone wires, giving a characteristic aspect to the vegetation. It is not a parasite, but merely roosts on the tree limbs and makes its own food as other plants do. The surface of the plant is covered with small, stalked scales, like

Fig. 22.43. Greenbrier (*Smilax hispida*), one of several species of *Smilax* which occur on the coastal plain. (Photo by John H. Gerard, from National Audubon Society)

Fig. 22.44. Venus's-flytrap, near Wilmington, North Carolina. (Photo by Walter Hodge)

little umbrellas. These are closely appressed to the surface in dry weather, but they loosen when wet and allow the plant to soak up rain or dew.

Thickets and hardwood forests in the coastal plain are often infested with one or more species of greenbrier (*Smilax*, Fig. 22.43), a tough, slender vine which usually has short, sharp prickles like cat's claws. There are both evergreen and deciduous species. Greenbriers are also found throughout most of the Eastern Deciduous Forest Province, but except toward the southern margin of the province they are not generally common enough to be a hazard to nature-lovers.

Asters and goldenrods are just as abundant and conspicuous in late summer and fall on the coastal plain as they are in the Eastern Deciduous Forest Province, but many of the species are different. Other members of the aster family are also abundant.

A great many of the species of the coastal plain are more or less confined to it, but others are widespread in the Eastern Deciduous Forest Province as well. A considerable number of the herbs are rare and local, or have very limited ranges. One of the most interesting of these is the Venus's-flytrap (*Dionaea muscipula*, Fig. 22.44), which is abundant in a limited area in North and South Carolina but is unknown elsewhere. Recently it has been frequently offered for sale as a house plant, but it is not easy to grow.

THE WEST INDIAN PROVINCE

The southern tip of Florida, from approximately Miami southward, belongs to the West Indian Floristic Province. The northern limit of the province is not at all precise, but several botanists have noted that many West Indian species extend north only to about latitude 26 or 27 degrees. The southern shore of Lake Okeechobee once supported considerable stands of tropical forest characterized by the custard apple (*Annona glabra*), a typical West Indian species, but this land is now largely under cultivation. Mangrove swamps, which are found on low tropical coasts all over the world, are well developed as far north as Naples on the Gulf Coast of Florida. On the Atlantic coast small patches of mangrove extend north nearly to St. Augustine, above 29 degrees latitude, but this is merely another example of the indefiniteness of most floristic boundaries.

The West Indian flora is a tropical flora, and its northward limits are determined largely by the incidence of frost. Many of the species can stand no frost at all, and these in general do not extend much if at all north of the

Fig. 22.45. Mangrove thicket in the Bahamas (Eleuthera Island). (Photo by Molly Adams, from National Audubon Society)

Florida Keys. Others can stand light frost, and these reach a greater or lesser distance farther north.

Many of the characteristic species of the Coastal Plain Province likewise extend southward into the West Indian province. The factors governing their southern limits are not as well understood as we might hope, but it appears that some of these species require a period of dormancy associated with cold weather, or at least their seeds require it for germination, whereas other species are simply unable to meet the competition of tropical species in a tropical climate.

Since we consider the West Indian Floristic Province from the standpoint of northerners, we probably draw the boundary farther north than we otherwise might. Traveling southward, we begin to find many new and unfamiliar species of West Indian affinities when we pass from the twenty-seventh to the twenty-sixth parallel of latitude, and we draw our line here. But if we were approaching the question from the standpoint of a West Indian botanist traveling northward, we might feel that so many northern species extend nearly or quite to the tip of Florida that practically the whole of mainland Florida should be referred to the Coastal Plain Province instead of to the West Indian Province.

Most of the natural vegetation of the West Indian Province as represented in Florida falls into four main types: coastal swamps, inland swamps, hammock forests, and the pine forests of the limestone ridge inland from the Atlantic coast. The coastal swamps are covered largely by mangroves; the inland swamps, such as the Everglades, are characterized by saw grass (*Mariscus jamaicensis*) and other coarse sedges and grasses; the hammocks carry a dense forest of many species of hardwoods; and the limestone ridge is characterized by slash pine (*Pinus caribaea*).

Low, muddy, tropical seacoasts the world over are likely to be covered with characteristic tangled trees called mangroves. Mangroves in cultivation do not require salt water (at least, the most common western hemisphere species does not), but under natural conditions they are generally unable to compete with other plants in fresh-water swamps. Several species of Mangrove belong to the genus *Rhizophora*, but others belong to genera in other families which are not closely related to the Rhizophoraceae. The most common and widespread mangrove in the New World, including southern Florida, is the red mangrove (*Rhizophora Mangle*).

Mangroves typically grow on tidal and subtidal flats, where the ground surface is covered with water for part or all of the day. They produce arching

Fig. 22.46. Saw grass marsh in the Everglades, with scattered hammocks in the background. (National Park Service photo)

Fig. 22.47. Gumbo limbo in a hammock on Big Pine Key, Florida. This characteristic West Indian tree also occurs in parts of Mexico, where it is known as the naked Indian (Indien desnudo) because of the bright reddish-brown bark. (Photo by Dade Thornton, from National Audubon Society)

prop roots, from which new aerial stems arise, so that a practically impenetrable thicket is formed (Fig. 22.45). Red mangrove commonly grows at the outer (seaward) margin of the thicket, where the surface is covered by water most or all of the time. Closer to the shore, where the surface is covered only at high tide, it is often replaced by the black mangrove (*Avicennia nitida*). The shoreward side of a mangrove swamp, near and above the normal high-tide line, supports various kinds of salt-tolerant plants, but farther out from solid ground there is usually little or nothing but mangrove.

Seeds of the red mangrove germinate while still on the tree, and the embryonic root may become a foot long before the seedling falls off. If the ground is exposed when the seed falls, the momentum of the fall effectively plants it in the soft muck. If the ground is covered, the seedling floats, root down, until the water recedes enough for it to get a foothold in the ooze.

The Everglades (Fig. 22.46) constitute a vast fresh-water marsh in southern Florida, with patches of open water and scattered "islands" of trees in a "sea" of coarse sedges, grasses, and rushes. The most common and characteristic plant, and over large areas almost the only obvious plant, is the saw grass (actually a kind of sedge). Low thickets of wax myrtle (*Myrica cerifera*) occur here and there. This is really a coastal plain species, which extends all the way north to New Jersey, rather than a West Indian species, and its abundance here serves to emphasize the transitional nature of the Everglades region.

The tree islands, occupying slightly drier ground in the Everglades, harbor a wide range of species. Some of these, such as the red bay (*Persea Borbonia*) and sweet bay are coastal plain species which approach their southern limit here. Others, such as gumbo limbo (*Bursera Simaruba*) and mastic (*Mastichodendron foetidissimum*) are West Indian species at their northern limit. These tree islands are essentially hammocks, as described in the following paragraphs.

The land is so flat in much of Florida that slight differences in elevation often have a disproportionate influence on the vegetation. Areas a few feet higher than their surroundings are commonly covered with hardwood trees. Such spots are called hammocks, the word being a variant of the more general and familiar term hummock. In central Florida the hammock trees are generally coastal plain species, but as one travels southward the coastal plain species are progressively replaced by West Indian ones. Even within the Everglades the composition of the hammock forests changes considerably from north to south.

The more southern hammocks support a dense tropical forest with many epiphytes, ferns, and lianas. Many different kinds of trees occur in these hammocks, reflecting the relative richness of tropical floras in general as compared to those of the temperate zone. Among the most noteworthy West Indian trees in the hammocks of southern Florida, in addition to several palms, are the gumbo limbo (Fig. 22.47), which attracts attention because of its smooth, coppery red bark, mahogany (*Swietenia Mahogani*, now largely destroyed by local lumbering), joewood (*Jacquinia keyensis*), wild cinnamon (*Canella Winteriana*), two species of fig (*Ficus aurea* and *Ficus brevifolia*), the wild dilly (*Manilkara emarginata*), mastic, and satinleaf (*Chrysophyllum olivaeforme*). These last three species belong to the Sapotaceae, an essentially tropical group which includes the species (*Manilkara zapotilla*) that is the prime source of the chicle used in chewing gum. All members of the family have a milky latex, but the properties of the latex differ in different species.

Many of the trees in the hammocks are evergreen, but some of them drop their leaves and are bare for several weeks or months before growing a crop of new ones. The six months from November through April are relatively dry, and the leaf fall is correlated with drought rather than with cold. In the tropics of both the Old and the New World, areas with an alternation of pronounced wet and dry seasons are often clothed with deciduous forest. The dry season in south Florida is not so extreme as in many other places, and a considerable number of the trees keep their leaves the whole year. Here, as in other areas where trees retain their leaves during a fairly dry season, evergreen trees have relatively firm, hard leaves which do not wither even when a shortage of water has forced them into dormancy.

The epiphytes in the hammocks belong to a number of different taxonomic groups, but the Bromeliaceae and Orchidaceae are particularly well represented. Some of them, particularly among the orchids, have very showy flowers, but others are relatively inconspicuous. The Spanish moss, which we have noted as a characteristic epiphyte of the Coastal Plain Province, is also well represented in the West Indian Province. It belongs to the family Bromeliaceae, but it is more highly modified than most members of the family.

Both of the common figs in the hammocks germinate as epiphytes, but do not remain so. After germinating they send down roots which reach the ground and then tend to coalesce laterally into either a stout trunk or a tube around the host tree. *Ficus aurea*, in particular, often encloses and eventually strangles its host, and is called strangler fig. Other species of strangler figs occur elsewhere in the tropics.

Not far inland from the Atlantic coast of southern Florida there is a limestone ridge, several miles wide, which rises a few feet above the surrounding

Fig. 22.48. Epiphytes on the trunk of a slash pine (*Pinus caribaea*) in southern Florida; the large specimen at the left is a bromeliad. (Photo by Walter S. Chansler, from National Audubon Society)

Fig. 22.49. Typical mixed hardwood forest in Puerto Rico. (U.S. Forest Service photo)

terrain. Where not cultivated or otherwise disturbed, this limestone ridge is generally covered by an open stand of slash pine (*Pinus caribaea*). The slash pine also extends northward for some distance into the Coastal Plain Province, as far as the coasts of Mississippi and Georgia. The distinction between the Coastal Plain and West Indian Provinces is least obvious in the pinelands, since the same general type of vegetation and many of the same species extend from one province into the other.

THE GRASSLAND PROVINCE

The Grassland Province occupies central North America, from the deciduous forests on the east to the Rocky Mountains on the west, from the coniferous forests of Canada on the north to the deserts of Mexico and southwestern United States on the south.

The climatic eastern limit of the province is not certain. The actual boundary was certainly shifted to the east by the repeated fires set by Indians bent on easy hunting. At least as far west as Manhattan, Kansas, native trees now do well in many sites and appear to be invading prairie land, even though they do not yet form a definite forest.

In pioneer days the eastern boundary of the grassland was a broad belt in which patches of prairie and forest formed a great mosaic. The forest extended into the grassland along all the streams and the grassland into the forest on the uplands. The individual boundaries between forest and grassland were commonly sharp, at least in the more northern areas, but the boundary between the two provinces as a whole was not.

Several thousand years ago, during what is called the postglacial xerothermic period, the climate in north temperate regions was somewhat warmer and/or drier than it is now. During that time the forest-prairie boundary was, of course, farther east than now, and a broad prairie peninsula extended across Illinois and Indiana into Ohio. Subsequently the forest reoccupied much of the land, and when the first white men came to the scene the prairie peninsula was broken up, toward the west, by patches of timber, and in its easternmost portion it was represented only by detached fragments of prairie on the driest sites. A few characteristic prairie plants still occur in isolated localities as far east as Ohio. Other vestiges of the prairie peninsula are the patches of black prairie soil, which does not develop under trees, now occupied by ordinary deciduous forest. The climatic basis of the prairie peninsula is shown even on modern meteorological maps. The more or less north-south lines of similar precipitation/evaporation ratios bulge eastward across Illinois and Indiana and into Ohio.

Grasses dominate the Grassland Province and give character to the landscape, but they are by no means the only plants. Various sorts of perennial and even annual herbs, many with showy flowers, occur in considerable abundance. The aspect of the vegetation changes from week to week and month to month during the growing season, as different genera and species come into flower and go on to set seed. Members of the aster family (Compositae) are prominent throughout the season, from the butterweeds (*Senecio*) and blanket-flowers (*Gaillardia*) in the spring, to the sunflowers (*Helianthus*) and coneflowers (*Rudbeckia* and *Ratibida*) of midsummer, to the asters and goldenrods of late summer and fall. Members of the pea family (Leguminosae) are also especially abundant in many places.

The Grassland Province is roughly divisible into three north-south strips, the tall-grass prairie to the east, the short-grass prairie to the west, and the mixed-grass prairie in between. The zonation reflects the increasing aridity from east to west. The term prairie, without further qualification, is often used to refer to the tall-grass prairie alone. The short-grass and mixed-grass country is then referred to as the plains, or the Great Plains.

The boundaries between these three types of prairie are irregular and unstable. In the drier, more exposed sites, the short-grass vegetation extends into the mixed-grass type and the mixed-grass into the tall-grass. In the swales, with a better moisture supply, the situation is reversed. A series of dry years shoves all boundaries eastward, and wet years push them west again.

Before the advent of the plowman the grasses of the tall-grass prairie formed a dense cover commonly 4 to 6 or even 10 feet tall, and, at least in the more favorable sites, a tough, thick sod was formed. Big bluestem (*Andropogon Gerardi*, Fig. 22.51,) little bluestem (*Andropogon scoparius*, Fig. 17.9), and Indian grass (*Sorghastrum nutans*) were among the commonest of the many dominant species, but there was a great deal of local and regional variation.

The deep, dark, fertile soil of the tall-grass prairie region is as good as any soil in the world, and the original vegetation has now been almost entirely replaced by productive farms, bustling cities, and other appurtenances of civilization. Much of the corn belt of the United States lies in the tall-grass prairie region, as do millions of acres devoted to wheat, cotton, and sorghum.

The high plains east of the Rocky Mountains, where the great herds of buffalo once roamed, are short-grass country. Buffalo grass (*Buchloe dactyloides*, Fig. 17.10) and little grama grass (*Bouteloua gracilis*, Fig. 17.11) are the most

Fig. 22.50. Bur oak in Buffalo County, Nebraska. In 1901, when this picture was taken, these trees had grown up from open prairie land in the course of 25 years. (U.S. Forest Service photo by W. L. Hall)

common species in relatively undisturbed sites. These and other common grasses in the region are generally only a few inches high, although taller species occur in the more favorable sites.

Rainfall in the short-grass region is scanty and irregular, commonly only 10 to 15 inches annually, and the later part of the summer is often very dry. The rain water seldom penetrates more than 2 or 3 feet into the soil, and a hardpan layer of accumulated calcium carbonate and other salts marks the level of penetration.

The short-grass prairie is best adapted to use as grazing land. In the more favorable years it is possible to grow wheat and other crops, but such years are inevitably followed by years of drought, crop failure, and dust storms. The great drought of the early 1930s, particularly 1934, put an end to most of the farming which the United States government had encouraged in this region as an emergency measure to boost food production during World War I. Put to the plow again during World War II, the land again produced a few good crops before the dry years returned; federal aid and the absence of a really severe drought have combined to permit continued cultivation of much of the short-grass region since that time, with mediocre success.

The mixed-grass prairie (Fig. 22.52) is, as its name implies, composed chiefly of elements from the tall-grass and short-grass prairies which flank it. The climate is favorable enough to permit many of the tall-grass species to survive (although they do not reach full size, seldom over 4 feet), but not favorable enough to permit them to crowd out the more drought-resistant, shorter species from farther west. The proportions of the various species fluctuate from year to year, principally according to the moisture conditions.

Most of the mixed-grass prairie region is suitable to the same crops as the tall-grass prairie, but the production is not so high and there is more chance of a crop failure during a dry year.

The Great Plains are not so flat as the name might imply. If one walked across them, or drove a covered wagon, he would have a far different impression of the terrain than he is likely to get by driving effortlessly along in an automobile. Elevated land marks are very few, but there are hills and swales, bluffs and steep slopes, as well as flatlands. The smaller streams are mostly intermittent rather than permanent, but they have nonetheless carved out valleys that often have sharply marked borders. North of the Missouri River the land has been glaciated, and the landscape reflects the uneven deposition of glacial till.

This diversity of habitats permits a more varied flora than might at first

Fig. 22.51. Big bluestem, one of the most characteristic species of the tall-grass prairies. (U.S. Conservation Service photo by Hermann Postlethwaite)

Fig. 22.52. Mixed-grass prairie in central Nebraska. This land is being used as a cattle range. Water is raised from a well by a wind mill. (U.S. Forest Service photo by Bluford W. Muir)

Fig. 22.53. Aspen-grass woodland near Vermilion, in southern Alberta. (Photo courtesy of J. Arthur Herrick)

be expected, but still nothing to compare with the provinces to the east, west, or south. After the tundra and the boreal forest, the Grassland Province is floristically the least diversified in the country.

Although the Grassland province is dominated by grasses, it is not completely treeless. There is a thin border of cottonwoods (*Populus deltoides* var. *occidentalis*, the Great Plains cottonwood) along most of the larger streams (Fig. 21.1). A dwarf phase of the bur oak (*Quercus macrocarpa* var. *depressa*) occurs wherever the moisture conditions are a bit more favorable than usual on the northern half of the plains. Usually hardly more than a shrub, it sometimes becomes as much as 15 feet tall. Other small trees occur here and there in favorable sites.

Several groups of wooded hills or low mountains rise out of the plains in the northern half of the Grassland Province. Among the most noteworthy of these are the Black Hills of South Dakota, the Turtle Mountains of northern North Dakota, and the Cypress Hills of southern Saskatchewan. The Black Hills are the largest and have attracted the most attention. Geologically, they are an eastern outlier of the Rocky Mountains. Botanically, they contain a mixture of cordilleran, boreal forest, and eastern deciduous forest species, as well as some species from the surrounding grasslands, but the cordilleran element is the most prominent. It would be defensible to treat the Black Hills as an enclave of the Cordilleran Forest Province in the Grassland Province. In this book we prefer to discuss such enclaves without giving them formal recognition on the map. Although these several groups of forested hills are very interesting to plant geographers, this interest stems from their combination of species from other provinces rather than from any individuality of their own.

Along the northern boundary of the Grassland Province there is a strip of aspen-grass woodland (Fig. 22.53), in which aspen occupies the north slopes and other protected habitats with grassland in between. In a classification of vegetation types, this might with some justification be treated as a separate type, as some botanists have done. It has no special flora of its own, however, and from the standpoint of floristic provinces it is no more than a transition zone between the Grassland and the Northern Conifer Province.

A few miles west of San Antonio in Texas the traveler will find one of the few 3-way transition zones to be seen in the country. In the swales there are a few live oak trees, the last vestige of the forests to the east. On the uplands are desert shrubs, underlain by a dense carpet of grass. To the north lie the grasslands of the Great Plains. South and west is the desert. If one travels the main east-west highway from this point, the oaks disappear and the grasses diminish to insignificance within a few miles to the west. Eastward the desert shrubs disappear and there are progressively larger patches of trees until one reaches a predominantly forest vegetation.

THE CORDILLERAN FOREST PROVINCE

The Cordilleran Forest Province occupies much of the area from the eastern base of the Rocky Mountains to the Pacific Ocean. At the north, in northwestern Canada, it merges with the Northern Conifer Province. Toward the south it is divided into two large prongs separated by the Great Basin Province. At its southernmost extreme, in New Mexico, it is broken into small bits on the higher mountains, isolated from each other by the more xeric vegetation of lower elevations. The Black Hills of South Dakota are another isolated fragment of the Cordilleran Forest Province, surrounded by grassland.

The Cordilleran Forest Province is dominated by coniferous forests. Broad-leaved trees—such as maple, birch, and (toward the south) oaks—also occur in the region, but the only one which is both widespread and abundant is the quaking aspen. Water is a critical factor in much of that part of the province that lies in the United States (Fig. 22.54), and toward the southern part of the province there is a considerable amount of open country, some of it grassland, some of it dominated by sagebrush or by piñon-juniper woodland, as in the Great Basin. Outside of Canada and the Pacific Northwest the forests seldom extend onto the lower valley floors between the mountains.

One of the most characteristic features of the Cordilleran Forest Province is the altitudinal zonation of different forest communities, so that one passes through several different associations as he climbs a mountain.

The climate in and west of the Cascade Mountains of Oregon and Washington is considerably moister than that of the Rocky Mountains farther east, and the two areas can conveniently be discussed separately. Some botanists prefer to treat them as representing two separate provinces, but they have much in common, and they fade into each other in northernmost United States and southern Canada, so that in a broad-scale view it is justifiable to treat them as one.

The Pacific Northwest, westward from the Cascade summits in Washington and Oregon, is the home of some of the tallest, most magnificent forests to be found anywhere in the world. On the Olympic peninsula of northwestern Washington they may fairly be said to form a temperate rain forest.

Fig. 22.54. Effect of moisture relations on vegetation in the mountains of northeastern Oregon. Coniferous forests on the relatively moist north slopes, blue-bunch wheatgrass (*Agropyron spicatum*) elsewhere. (U.S. Forest Service photo by Melvin H. Burke)

From the lowlands to moderate elevations in the mountains of western Washington and Oregon the most common trees are Douglas fir (*Pseudotsuga Menziesii*, Fig. 22.55), western hemlock (*Tsuga heterophylla*), and western red cedar (*Thuja plicata*, Fig. 22.56). Over much of this area, except the western side of the Olympic Mts., the summers tend to be rather dry, and since prehistoric times there have been sporadic forest fires. Douglas fir has a very thick bark and is relatively resistant to fire, as compared to the thin-barked hemlock and cedar. Its seeds germinate best in mineral soil from which the surface litter has been burned off or otherwise removed. Fire therefore tends to favor Douglas fir over the other two species. On the other hand, seedlings and young trees of Douglas fir are much less tolerant of shade than those of cedar and hemlock. The seeds of these latter two species are well adapted to germinating in a soil rich in organic matter. Absence of fire therefore tends to favor cedar and hemlock. Thus we have the paradox that the magnificent stands of Douglas fir, one of the most valuable timber trees in the world, are in the long run, under natural conditions, dependent on fire. The balance is a delicate one, however, and there are now so many man-caused fires that every effort must be made at fire-control in order to maintain any forests at all. It also seems likely that with proper forest management it will be possible to grow repeated crops of Douglas fir on the same land without the intervention of fire.

At upper elevations in the mountains the Douglas fir, cedar, and western hemlock give way to other conifers. One of the commonest trees in this highland forest is the mountain hemlock (*Tsuga Mertensiana*). Western white pine (*Pinus monticola*), which forms extensive forests in the mountains of northern Idaho (Fig. 22.57), is also a frequent tree at upper altitudes in the Cascade Mountains.

The redwood forest (Fig. 20.1) in the coast ranges of northern California and extreme southwestern Oregon has been mentioned several times in earlier chapters. These are the tallest trees in the world, reaching a height of more than 300 feet. Dependent more on fog than on rain, they form the southwestermost finger of the Cordilleran Forest Province.

The coast ranges of Washington, Oregon, and California are in general lower than the more inland Cascades and Sierra Nevada, but they do cast a minor rain shadow over the Puget trough. A gradually attenuating tongue of Mediterranean climate extends northward from California between the Cascade and Coast ranges. At Portland, Oregon, for example, in the Willamette Valley, it seldom rains during June, July, and August. The climate

Fig. 22.55. Dense forest of Douglas fir on the west side of the Cascade Mountains in Pierce County, Washington. The small trees coming up under the Douglas fir can be recognized as hemlocks by the fact that the leader is nodding instead of stiffly erect. (U.S. Forest Service photo by A. Gaskill)

is of course reflected in the vegetation. There were bits of natural prairie in the Willamette Valley and about Puget Sound, and the madrone (*Arbutus Menziesii*), a characteristic tree of the broad sclerophyll forests of California, extends all the way north to Puget Sound and the southern tip of Vancouver Island.

The summer drought is scarcely noticeable in western British Columbia, southern Alaska, and on the west side of the Olympic Mountains in Washington. Here the Sitka spruce (*Picea sitchensis*), another very large tree, becomes a dominant member of the community, along with western hemlock. Ferns grow luxuriantly in this spruce-hemlock forest (Fig. 22.58), and the ground is often covered with moss. Devil's club (*Oplopanax horridum*) is a common shrub in this community. Its stem is beset with stout, sharp prickles, and the climber who grasps it while struggling up a timbered slope is not likely to do so a second time.

Returning to Oregon and Washington, we find that the east slope of the Cascade Mountains is much drier than the west slope, and there are correlated differences in vegetation. In Oregon ponderosa pine is the most characteristic tree of the east slope of the Cascades (Fig. 22.61). We shall see that ponderosa pine is also common in other fairly dry parts of the Cordilleran Forest Province.

The Sierra Nevada is drier than the Cascades, and the coniferous forests are confined to progressively higher altitudes toward the south. Indeed the aspect of the forest is often more like that of the Rocky Mountains, which are also relatively dry, than like that of western Washington and Oregon. Ponderosa pine is a common species toward the lower margin of the forest in the Sierra Nevada. At middle altitudes this gives way to the closely related Jeffrey pine (*Pinus Jeffreyi*), which is often associated with red fir (*Abies magnifica*, Fig. 22.59). The sugar pine (*Pinus Lambertiana*, Fig. 22.60), with cones up to 2 feet long, occurs in both of these zones in the northern Sierra Nevada. Higher up in the mountains the most characteristic tree is the mountain hemlock. In the Sierra Nevada, as in the Cascades, this is often associated with one or more species of five-needled pines. These species extend up to timberline. Above that is the arctic-alpine vegetation, representing detached fragments of the Tundra Province.

The bigtree groves of the Sierra Nevada are of course well known. They occur at middle altitudes, commonly about 6,000 feet. Although not quite so tall as the coast redwood, the sierra redwood (i.e., the bigtree) is a good deal bulkier. Reaching a diameter of more than 25 feet, they are by all odds the

Fig. 22.56. Mature stand of western red cedar on the west side of the Cascade Mountains in northern Washington. The shreddy bark can be peeled from the trunk in long, vertical strips. (U.S. Forest Service photo by E. Lindsay)

Fig. 22.57. Virgin forest of western white pine, 300 years old, in northern Idaho. (U.S. Forest Service photo by K. D. Swan)

Fig. 22.58. Spruce-hemlock forest in southwestern British Columbia. The jumble of undergrowth and down timber is typical. The ground here is deeply covered with moss which never dries out. (U.S. Forest Service photo by E. B. McKay)

Fig. 22.59. Red fir forest in the Sierra Nevada of California. Here, as in other dense coniferous forests, the lower branches are shaded out and die as the trees get taller, but often they remain on the tree for some years after dying. (U.S. Forest Service photo)

Fig. 22.60. Sugar pine in Lassen Volcanic National Park, California. The wide, irregular crown, with cones near the ends of the upper branches, is typical. (Photo by John W. Sumner, from National Audubon Society)

Fig. 22.61. Ponderosa pine forest at the eastern foot of the Cascade Mountains in central Oregon. This is climax forest, and the young trees coming up are also ponderosa pine. (U.S. Forest Service photo by Melvin H. Burke)

most massive of all living things. They have very few natural enemies, and they are rarely killed even by fire. The author has seen a burned out shell of a tree, which one could step into and look out at the sky. The bark had all been burned away at the base except for a strip about 18 inches wide. Yet the tree still had live branches near the top on that side, and the connecting strip of bark was vigorously expanding its width. The species sets abundant seed, and young trees as well as old ones can be seen in many of the groves. With all these things in its favor, it should, seemingly, be much more widespread and abundant than it is, but apparently it has a very rigid set of requirements in order to grow well and compete with other trees. One of the things it may need is a thick blanket of snow as a protection of the soil against winter cold. In the northeastern United States, individuals may do well for several years or even decades, only to die during some winter when there is a severe cold spell without much snow on the ground.

The most characteristic tree toward the lower limit of the forest in the Rocky Mountain region of the United States (and west to the Cascade-Sierra summits) is the ponderosa pine (Fig. 22.61). Along the eastern front of the mountains, open forests of ponderosa pine merge with the prairie grasslands. Farther west they abut on sagebrush, or on piñon-juniper desert forests of the Great Basin type, or onto local grasslands, such as those of the Palouse region (Fig. 22.62) in southeastern Washington.

The highest forest zone in the Rocky Mountains, well separated from the ponderosa pine, is usually characterized by spruce and fir, with relatively small, narrow, spire-topped trees (Fig. 22.63). The fir is mostly alpine fir (*Abies lasiocarpa*). Toward the north the Spruce is Engelmann spruce (*Picea Engelmannii*); farther south, notably in Colorado, it is the closely related Colorado blue spruce (*Picea pungens*). As in the Cascade-Sierran region, five-needled pines are also found at upper altitudes, especially near timber line. Limber pine (*Pinus flexilis*) and whitebark pine (*Pinus albicaulis*) are characteristic members of this group.

At middle elevations, between the ponderosa pine and the spruce-fir zone, the most common dominant tree is the Douglas fir (Figs. 22.64, 22.65). The Rocky Mountain form of Douglas fir is smaller than the coast form, and it is adapted to a colder, drier climate. It sometimes extends all the way to timber line, and in areas where ponderosa pine is missing it may adjoin the sagebrush or the juniper slopes of lower altitudes.

Lodgepole pine (*Pinus contorta* var. *latifolia*) is also frequent at middle elevations in the Rocky Mountains, and less commonly in the Cascades and

Fig. 22.62. Palouse prairie near Pomeroy, in southeastern Washington. The deep, aeolian soil of the Palouse region is highly fertile, and most of the land is now covered with productive fields of wheat. Under natural conditions the grasses, notably blue-bunch wheatgrass, form a dense turf. (U.S. Forest Service photo by G. D. Pickford)

Fig. 22.63. Subalpine forest near Henry's Lake in Idaho, looking toward the Madison Range in Montana. The trees here are alpine fir, Engelmann spruce, and whitebark pine. (U.S. Forest Service photo by A. T. Boisen)

Fig. 22.64. Typical Douglas fir tree in Colorado. The forest here is more open, and the trees smaller, than in the Douglas fir forests west of the Cascade summits in Oregon and Washington. (U.S. Forest Service photo by E. S. Shipp)

Fig. 22.65. Twig and cone of Douglas fir. The cones are easily recognized by the long, three-parted bracts emerging from between the cone-scales. The needles are flat, blunt, and twisted at the narrow base, and the terminal bud is rather sharply pointed. (U.S. Forest Service photo by W. D. Brush)

Fig. 22.66. Dense stand of lodgepole pine near Neihart, Montana. (U.S. Forest Service photo by K. D. Swan)

Fig. 22.67. Lupine (*Lupinus polyphyllus*) near Leavenworth, Washington. Many other species of lupine occur in western United States. (U.S. Forest Service photo)

Fig. 22.68. Indian paintbrush (*Castilleja*) in the Cascade Mountains of Washington. Paintbrushes get some of their nourishment from the roots of other plants which they parasitize, and the leaves are often rather pale green instead of bright green, reflecting a low concentration of the green pigment (chlorophyll) which is essential for the manufacture of food. (U.S. Forest Service photo by Lloyd F. Ryan)

Sierra Nevada. It often grows in dense stands of small, very slender trees (Fig. 22.66) with live branches only toward the top. It was therefore favored by the Indians for use as lodge poles. Lodgepole pine is typically a fire-tree, much like the closely related jack pine of the boreal forest. In a few areas, such as Yellowstone Park, lodgepole pine approaches the status of a climax tree which can perpetuate itself over successive generations without fire. Douglas fir, which in the Rocky Mountains is a climax tree, often replaces lodgepole pine after some years.

Aspen groves (Fig. 22.80) are another characteristic feature of middle and upper elevations in the Rocky Mountain (and Great Basin) region. As we have already noted, the tree spreads mainly underground, and seedlings are rare. The margins of a grove expand into the adjacent, unforested area during relatively moist years, and the peripheral trees are killed back during years of drought.

In the Rocky Mountains, as elsewhere in the Cordilleran Forest Province, the highest reaches of the mountains are treeless and carry bits of alpine-tundra vegetation.

The less densely forested parts of the Cordilleran Forest Province have many showy perennial wild flowers, including asters, beardtongues (*Penstemon*), buttercups, daisies, larkspurs (*Delphinium*), lupines (Fig. 22.67), paintbrushes (*Castilleja*, Fig. 22.68), and shooting stars (*Dodecatheon*). There are but few annuals. In the deeper forests the flowers are fewer, and these flowers include a number of members of the heath and orchid families, as in the Northern Conifer Province.

THE GREAT BASIN PROVINCE

The hydrographic Great Basin, which has no drainage to the sea, makes up the core of the Great Basin Province. Western Utah, nearly all of Nevada, and small parts of eastern California, southeastern Oregon, southern Idaho, and southwestern Wyoming belong to the hydrographic Great Basin. Much of the more eastern part of the Great Basin drains into the Great Salt Lake in Utah, but there are other interior drainage basins of varying size in the Great Basin as a whole. Some of these have shallow, permanent salt lakes which vary in size according to the season. Many others have playa lakes that contain water after the spring run-off but dry up in summer, leaving a barren alkaline playa the rest of the year. And many streams come down from the mountains and disappear into the desert without reaching even a temporary lake.

The whole western margin and parts of the northern and northeastern

Fig. 22.69. Sagebrush-grass community on the Snake River Plains in Idaho. South slope of Big Butte at the upper left. Mile after mile of the Snake River Plains looks much like this, or differs mainly in having less grass. (U.S. Forest Service photo by Joseph F. Pechanec)

Fig. 22.70. Blue-bunch wheatgrass and scattered sagebrush on an isolated, cliff-sided plateau at the junction of the Deschutes and Crooked Rivers in central Oregon. Much of the more northern part of the Great Basin Province probably looked much like this a century ago, before the competitive balance was shifted by heavy grazing. (U.S. Forest Service photo by Richard S. Driscoll)

margins of the hydrographic Great Basin abut against the Cordilleran Forest Province. A narrow marginal strip of the hydrographic Great Basin must be included in this other province, as for example the eastern slopes of the Sierra Nevada and Cascade Mountains in California and southern Oregon.

In addition to the bulk of the hydrographic Great Basin, the Great Basin Floristic Province includes some adjacent areas which drain to the sea through the Colorado and Columbia Rivers. Almost all of eastern Utah and northern Arizona drains into the Colorado River, but except for the Uinta Mountains this area floristically belongs to the Great Basin. The Snake River plains of southern Idaho (Fig. 22.69) likewise have a Great Basin Flora, and a progressively attenuated element of the Great Basin flora extends north through eastern Oregon into eastern Washington and even into the Okanogan Valley in southern British Columbia.

Since much of the Great Basin Province has no drainage to the sea, the mineral salts which would otherwise be carried away to the ocean are carried only a relatively short distance from their source in the native rocks, and they tend to accumulate in places where water evaporates. Even in the areas which nominally drain into the Colorado and Columbia Rivers, there are many local basins which drain poorly or not at all. Soil with an oversupply of mineral salts is said to be alkaline. In the more extreme sites these salts form a thin white crust (alkali) on the surface of the ground.

The Great Basin Province is cool desert country, with forests only in some of the higher mountains. The Cascade-Sierran axis casts a great rain-shadow for hundreds of miles to the east, and the Rocky Mountains catch most of the moisture that might come in from the Gulf of Mexico. Between the Cascade-Sierran axis and the Rocky Mountains the air is generally dry. What precipitation there is comes mostly in the winter, although there are summer thunder-showers especially in the mountains. In the valleys the annual precipitation is often less than 10 inches, and seldom more than 15. Along the Snake River where it forms the boundary between Oregon and Idaho the elevation is only about a thousand feet, but in most of the province the valleys are at three to six thousand feet above sea level.

The most characteristic plants of lower elevations in the Great Basin Province are sagebrush (*Artemisia tridentata*, Fig. 17.2), greasewood (*Sarcobatus vermiculatus*), and shadscale (several species of *Atriplex*). Sagebrush is by all odds the most abundant. They are all shrubs, seldom as tall as a man, with small leaves and inconspicuous, small flowers. The leaves of sagebrush and shadscale are gray-green rather than bright green, and much of the desert therefore has a grayish aspect.

Fig. 22.71. Alkaline flats dominated by shadscale and winter fat in southwestern Utah. The paler plants are winter fat; the darker, often larger ones are shadscale. (U.S. Forest Service photo by Selar S. Hutchings)

Fig. 22.72. Luxuriant stand of halogeton on overgrazed shadscale range in southern Idaho. Halogeton, a salt-tolerant member of the beet family, native to interior Asia, has been introduced in our arid west and is becoming progressively more common. It is poisonous to livestock. (U.S. Forest Service photo by Selar S. Hutchings)

Fig. 22.73. Arrow-leaved balsamroot in southeastern Oregon. This is one of the most common and showy spring flowers in the northern part of the Great Basin Province and northward in the drier parts of the Cordilleran Forest Province. (U.S. Forest Service photo by L. D. Bailey)

Fig. 22.74. Sego lily, the state flower of Utah, growing among sagebrush in central Utah. (U.S. Forest Service photo by Lincoln Ellison)

Fig. 22.75. Typical Utah juniper near Delta, Utah.

Fig. 22.76. Typical single-leaf piñon, near Wheeler Peak in east-central Nevada.

Fig. 22.77. Piñon-juniper community in central Utah. The forest here is more open than in many places, with much bare, rocky ground. (U.S. Forest Service photo by A. R. Croft)

Fig. 22.78. Piñon-juniper community near Flagstaff, Arizona. The forest here is denser and taller than in many other places. (U.S. Forest Service photo)

Sagebrush is pollinated by wind, and some people react to the pollen. During the late summer and early fall, when many people in the eastern and central United States are suffering from hay fever caused by ragweed pollen, some people in the Great Basin have similar problems with sagebrush pollen.

Sagebrush is sensitive to alkali; it will not grow where mineral salts accumulate in the soil. It grows much taller in deep, fertile soils than in shallow ones, partly because of the response of individual plants to varying conditions, and partly because of natural selection of different types for different habitats. Where the sagebrush is more than waist-high the soil will probably produce good crops under irrigation. Where it is less than knee-high, the land is probably not worth farming even if water is available.

Grasses and other herbs occur in varying amounts with the sagebrush. Especially noteworthy are two perennial bunch-grasses, the blue-bunch wheatgrass (*Agropyron spicatum*) and sheep fescue (*Festuca ovina*); these are likely to be found wherever there is sagebrush. In some areas, especially toward the northern part of the province, the grasses were probably more abundant and conspicuous in pioneer days than the sagebrush (Fig. 22.70). Heavy grazing has caused a considerable diminution of the grasses over much of the area, until in some places there is no native grass left.

Sagebrush is only one of many plants which do not tolerate alkali. The beet family (Chenopodiaceae) includes an unusually high percentage of species which do tolerate alkali, and members of this family are especially common in alkaline soils. Both greasewood and shadscale belong to the beet family.

Where the soil is too alkaline for sagebrush, greasewood and shadscale often dominate the scene (Fig. 22.71). Winter fat (*Eurotia lanata*), another gray-leaved member of the Chenopodiaceae, is also common in such sites. The plant gets its name because it furnishes excellent, nutritious forage for sheep in the winter. It would doubtless be acceptable in the summer too, but then the sheep are enjoying the more luxuriant vegetation in the mountains.

June grass (*Bromus tectorum*), a winter annual introduced from southern Europe, is now abundant in the desert country throughout most of the Great Basin Province. It comes up in the fall, lies dormant through the winter, and resumes growth in the spring. By June it is about a foot high and mature or nearly so. For the rest of the summer it is tinder-dry, and its long-awned spikelets easily work their way into the clothing of anyone who wanders by. After it begins to mature, June grass is generally avoided by grazing animals, because of the awns, so the seeds remain to germinate during the fall rains.

There is a good display of perennial flowers in the spring and early summer

Fig. 22.79. Mountain mahogany, surrounded by sagebrush, at the western edge of the Great Basin, Devil's Gate Pass, California.

Fig. 22.80. Aspen grove in southwestern Montana, with sagebrush on the open slope at the right. Similar groves are common in the Great Basin Province. (U.S. Forest Service photo by K. D. Swan)

in the sagebrush regions, although it does not compare with the spring flush of annuals in the Sonoran desert to the south. One of the most conspicuous of the perennials, in the northern part of the province, is the arrow-leaved balsamroot (*Balsamorhiza sagittata*, Fig. 22.73), a coarse low plant with sun-flower-like flower-heads and large, gray-green, arrowhead-shaped leaves. The sego lily (*Calochortus Nuttallii*, Fig. 22.74), the bulbs of which served the Mormon pioneers as emergency food, is another common and conspicuous wild flower. The showy Indian paintbrush (*Castilleja*) is represented by many species with bright red to yellow bracts among the flowers, and there are even more numerous species of beardtongue, these most often with bright blue flowers. Lupines (*Lupinus*) and several low species of phlox are also common and showy. Most numerous of all, in numbers of species, are the locoweeds (*Astragalus*), with flowers like small sweet peas, but many of these have relatively small and inconspicuous flowers, and the individuals of each kind are not always abundant. Some of the species of *Astragalus* have very limited natural ranges, only a few miles across, and they are often restricted to particular habitats, such as those rich in selenium. All of these several genera of wild flowers are well-represented in the more open areas in the Cordilleran Forest Province, as well as in the Great Basin Province.

There are also annual spring flowers in the sagebrush desert, notably some species of *Phacelia*, but many of the annuals have relatively small flowers, and they seldom form massive displays.

The Great Basin is not simply a bowl surrounded by mountains. It has many mountain ranges of its own, nearly all narrow and trending more or less north-south. As one climbs from the valley floors toward the mountain tops, the annual precipitation (both rain and snow) increases, and the average temperature decreases, so that the evaporating power of the sun is lessened. Increasing altitude is thus associated with a progressive shift in the ratio of precipitation to potential evaporation (the *p/e* ratio). In the Great Basin region, as well as in the Cordilleran Forest Region, we therefore have a pronounced vertical zonation of the vegetation.

The foothills and lower slopes of the mountains in the Great Basin Province often support an open, low, desert forest of junipers, especially Utah juniper (*Juniperus osteosperma*, Fig. 22.75), or at slightly higher elevations the Rocky Mountain juniper (*Juniperus scopulorum*). In Utah, Nevada, and northern Arizona the junipers are often mixed with one or another of two species of nut-pine, the single-leaf piñon (*Pinus monophylla*, Fig. 22.76), and the two-leaved piñon (*Pinus edulis*). These pines have relatively large seeds which were used as food by the Indians. The seeds can often be bought locally as piñon-nuts. In Utah, scrub oaks (*Quercus utahensis* and other species) often form dense thickets in habitats comparable to those occupied by piñon and juniper. Both the oak and the piñon extend as far north as Cache Valley in northern Utah, but neither is known to enter Idaho.

One of the most interesting plants of this dry forest zone above the valley floors is the mountain mahogany (*Cercocarpus ledifolius*, Fig. 22.79). No relation to the true mahogany of the American tropics, mountain mahogany is a member of the rose family. It is a slow-growing, coarse shrub or small tree, seldom over 12 feet tall, with small leaves and very hard, dense wood. It sinks in water, and it is best cut with a dull axe, since it may chip a sharp one. It is of no great economic importance, but it is sometimes used locally as firewood, burning with a very hot, clean flame. The author has seen a pickup truck resting springily on top of a mountain mahogany thicket on a mountainside just below a narrow road, after failure of the steering gear.

The grasses and broad-leaved herbs of this desert forest in the Great Basin region mostly belong to the same genera as those of the sagebrush, but the species are often different.

At higher elevations in the mountains of the Great Basin region, above the zone of desert forest, there are often coniferous forests and aspen groves (Fig. 22.80) much like those of the Rocky Mountains, and many of the same or similar wild flowers also occur. Some of the highest mountains, such as the La Sals in southeastern Utah, extend well above timber line and carry fragments of alpine tundra.

THE CALIFORNIAN PROVINCE

Most of California, from the Pacific Ocean to and including the western foothills of both the Sierra Nevada and the Cascade Mountains, belongs to the Californian Province. In the northern coast ranges (north of San Francisco Bay) and in the mountains of northern California and southwestern Oregon, the Californian Province interfingers extensively with the Cordilleran Forest Province. A progressively attenuated Californian element extends northward through Oregon, between the Cascade and Coast Ranges, and reaches even to Puget Sound in Washington. In southeastern California the chaparral of the Californian Province gives way to the desert of the Sonoran Province. At middle altitudes in the mountains and foothills of Arizona and New Mexico the aspect of the vegetation is often similar to that of California, and these areas are here for convenience discussed under the Californian Province.

Fig. 22.81. White columbine (*Aquilegia caerulea* var. *ochroleuca*) in the Wasatch Mountains of Utah. Typical *Aquilegia caerulea*, with blue and white rather than wholly white flowers, is more common in the Rocky Mountains proper and is the state flower of Colorado. (U.S. Forest Service photo by A. G. Nord)

Fig. 22.82. Chaparral-clad hills in southern California. (U.S. Forest Service photo by C. Miller)

The Californian Province characteristically has a Mediterranean type of climate, as discussed in an earlier chapter, and the vegetation, as always, reflects the climate. Water is a critical factor throughout the province, and slight differences in moisture balances can cause striking differences in the vegetation.

The topography and soils of the Californian Province are highly diversified, and there is a bewildering complexity of microhabitats. Although there are often only a few species in each particular habitat, there are so many different habitats, each with its own array of species, that the province has a very rich flora. All this is in striking contrast to the situation in some other provinces, such as the Tundra and the Northern Conifer Province, in which the species are much fewer but more ubiquitous.

Chaparral and open oak-woodland are the commonest general types of communities in the Californian Province. In the moister or more protected sites the community becomes a real forest composed mostly of broad-leaved evergreens. In the drier places, as in the great Central Valley of California, it is often grassland.

The chaparral community (Fig. 22.82) is dominated by evergreen shrubs mostly from 3 to 10 feet tall, with firm, mostly rather small leaves—broad sclerophylls, in ecological terminology. These shrubs typically have very rigid, crooked branches, and the individual plants are often closely set, forming an almost inpenetrable thicket. Chamise (*Adenostoma fasciculatum*) and various species of manzanita (*Arctostaphylos*) and "California Lilac" (*Ceanothus*, Fig. 22.83) are characteristic dominants of the chaparral community. A species of scrub oak (*Quercus dumosa*) is another common shrub in the chaparral.

The chaparral community is subject to frequent fires. This has evidently been so for thousands of years. The most successful species of the community are not generally killed by fire, even though they may burn down to the ground. After the fire they send up new sprouts from near the ground line, and within a few years the vegetation is back to normal.

The oak woodlands (Fig. 22.84) of the Californian Province occur in habitats not obviously very different from that of the chaparral, although the climate is often a little cooler or moister. The individual trees range from 20 to 75 feet tall at maturity and are usually well scattered, so that the aspect is frequently that of a savanna rather than a forest. Sometimes, on the other hand, the oaks form dense thickets reminiscent of the chaparral.

Several different species of oaks occur in the oak woodlands, some of them deciduous, others evergreen. Perhaps the most characteristic species are *Quercus Wislizenii*, the interior live oak, and *Quercus chrysolepis*, the canyon oak, which extends all the way east to New Mexico. Both of these species are evergreen. Among the deciduous species, *Quercus Kelloggii*, the California black oak, *Quercus lobata*, the valley oak, and *Quercus Garryana*, the Oregon oak, are probably the most common.

Patches of oak woodland are frequent in the drier sites of southwestern Oregon, where the moister areas are covered by coniferous forest of the Cordilleran Forest Province. Northward from Eugene the oak woodland is progressively submerged in the coniferous forest and scarcely forms a distinctive community, although one species, the Oregon oak, extends all the way north to Puget Sound in Washington.

In the inner coast ranges and on the western foothills of the Sierra Nevada, the oak woodland often gives way to an open forest of digger pine (*Pinus Sabiniana*), sometimes accompanied by Coulter pine (*Pinus Coulteri*). These are openly branched, rather small trees, often without a central axis in the upper part. They have massive, strongly armed cones with large, edible seeds. At its upper altitudinal limit this forest often gives way to forests of ponderosa pine. The digger pine forest stands in about the same ecological relationship to the montane coniferous forests of California as the desert forest of piñon-juniper in the Great Basin region does to the montane coniferous forests of that area.

The western side of the coast ranges in California harbors several species of pines and cypresses that have very limited natural ranges. The most famous of these are the Monterey pine (*Pinus radiata*) and the Monterey cypress (*Cupressus macrocarpa*, Fig. 22.85), the latter known only from the Monterey peninsula. These and several other species occur on exposed slopes with relatively cool summers and considerable fog. To the north they occur on the seaward side of the redwoods, in more barren soils. The bishop pine (*Pinus muricata*), which ranges from Humboldt County to Santa Barbara County, is the most characteristic species of this community, which has sometimes been called the closed-cone pine forest. The closed-cone pine forest is not continuous, but forms small patches here and there near the coast.

The more favored sites in the Californian Province, especially in the coast ranges from San Francisco northward, often support a broad sclerophyll (evergreen) forest. Some of the oaks from the oak woodland, notably the canyon oak, enter this forest, but four other trees, all of which reach heights of more than 100 feet, are more characteristic. These are the giant chinquapin (*Castanopsis chrysophylla*), the tanbark oak (*Lithocarpus densiflora*), the California laurel (*Umbellularia californica*), and the madrone (*Arbutus Menziesii*). The

Fig. 22.83. California lilac (*Ceanothus crassifolius*) in bloom in southern California. (U.S. Forest Service photo by C. C. Buck)

Fig. 22.84. Open oak woodland in the foothills of the Sierra Nevada, along the Kaweah River, in California. (U.S. Forest Service photo by A. Gaskill)

Fig. 22.85. Monterey cypress trees at Cypress Point on the Monterey Peninsula, California. (U.S. Forest Service photo by C. W. Johnson)

Fig. 22.86. Grassland dominated by wild oats, in the valley near the southwestern base of the Sierra Nevada of California, in 1909. (U.S. Forest Service photo by A. W. Sampson)

madrone is especially conspicuous because the outer bark peels irregularly from the trunk, exposing the shining, reddish brown inner bark. All of these species can grow individually in the coniferous forests also, and the broad sclerophyll forest passes directly into the coniferous forest, as well as into the chaparral and the oak woodland. The California laurel, in fact, reaches its best development in southwestern Oregon, where it grows along streams, intermingled with conifers. There it is known as the Oregon myrtle, and local people will tell you that it grows only "here and in the Holy Land."

The hotter and drier sites in the Californian Province, especially on the floor of the Central Valley, were originally occupied mainly by grassland. Various perennial bunch grasses dominated the scene. Much of this area is now cultivated under irrigation, or has homes built upon it, and the original vegetation has largely disappeared. What remains has been so heavily grazed that the native grasses are mostly gone, replaced by weedy annual grasses introduced from Europe. The slender wild oats (*Avena barbata*) is one of the commonest of these weedy grasses in California.

This California grassland does not fit the pattern of low winter and adequate summer precipitation that was explained in Chapter 20 as being typical of grasslands in temperate climates. Here the precipitation comes mainly in the winter, and the summers are very dry. In terms of the seasons of growth and dormancy, however, rather than of winter and summer as such, the correspondence to pattern is very good. The climate is warm enough to permit plants to grow during much of the winter, and the hot dry summers are the dormant season. Thus here, as in other grasslands, we have a pattern of reasonably adequate moisture during the growing season, combined with drought during the dormant season.

One of the most noteworthy special habitats in the Californian Province is that provided by outcrops of serpentine rock. Serpentine is a metamorphosed form of peridotite, an ultrabasic lava which is very low in available calcium. Serpentine soils are relatively infertile and are inhospitable to many kinds of plants. Both the superabundance of some minerals and the scarcity of others contribute to its infertility, but the insufficiency of calcium is believed to be the most important factor.

Many species are wholly restricted to serpentine soils in nature, sometimes to a single outcrop only a few acres in extent. One of the best places in the United States to look for previously unknown species of flowering plants, species that are "new to science," is on serpentine. Every individual outcrop that has not already been carefully studied holds the potentiality of bearing a "new" species, though it must of course be added that this potentiality will only occasionally be realized.

Serpentine outcrops are scattered over a good share of northern California, especially in the coast ranges, but the largest deposits are in southwestern Oregon. These latter deposits are in an area called the Klamath region, in southwestern Oregon and northwestern California, which is a floristic subprovince of its own. On the basis of the dominant woody species, the Klamath region is mainly transitional between the Californian and Cordilleran Forest Provinces, without any individuality of its own, but on the basis of the herbaceous plants it is a well-marked unit with many endemic species, both on and off the serpentine rocks.

Many showy species of both annual and perennial wild flowers abound in the Californian Province. One of the best known is the California poppy (*Eschscholzia californica*). Members of the aster family (Compositae), the phlox family (Polemoniaceae), the waterleaf family (Hydrophyllaceae), the evening primrose family (Onagraceae), and the lily family (Liliaceae) are particularly common, but the species and even many of the genera are different from those commonly encountered outside the province.

The Californian Province has the most sharply differentiated flora in the nation. No matter what other provinces a visitor may be acquainted with, he will find but few familiar plants here. The families are in general the same as elsewhere in the country, but many of the genera and especially the species are not found much if at all beyond the borders of the state. A visitor from the Rocky Mountains, or the Great Basin, or the Pacific Northwest will feel reasonably at home in the Sierra Nevada, but let him descend even to the foothills on the western side and he is lost. He will recognize the oaks and pines as oaks and pines, but he will probably not know the species, and many of the genera of herbaceous plants will be wholly unfamiliar.

Chaparral and oak woodlands resembling those of California also occur in parts of Arizona (Fig. 22.87) and New Mexico, especially in bands encircling the isolated mountain ranges of central to southeastern Arizona and southwestern New Mexico. The resemblance is mainly in broad aspect and in some of the dominant genera, however. Many of the other genera and most of the species are different. *Quercus turbinella*, *Arctostaphylos pungens*, and *Ceanothus Greggii* are common members of the Arizona-New Mexico chaparral which *do* extend westward to southern California. The general affinities of the flora, however, are with the Sonoran and to some extent also the Great Basin and Cordilleran Forest Provinces, rather than with the Californian Province.

Fig. 22.87. Open oak woodland in the Santa Rita Mountains of southern Arizona. The Oak here is *Quercus Emoryi*, which occurs from western Texas to Arizona and northern Mexico but does not reach California. (U.S. Forest Service photo by Kenneth W. Parker)

THE SONORAN PROVINCE

The warm desert country of southwestern United States belongs to the Sonoran Province, which also extends well down into Mexico. Southeastern California, southernmost Nevada, most of southern and western Arizona, most of southern New Mexico and Texas, and a tiny corner of southwestern Utah belong to this province.

Like the cool desert of the Great Basin Province, the warm desert of the Sonoran Province is interrupted by scattered mountain ranges which sometimes rise high enough to carry coniferous forests or even alpine tundra. Within the United States the warm desert is confined to lower elevations, mostly less than 3,000 feet; some of it, such as Death Valley in California, is below sea level.

Moisture is even more deficient and uncertain in the warm desert than in the cool desert. In the United States the annual precipitation in the warm desert is often less than 5 inches. The summers are long and hot, and the winters are mild. Part of the area does have occasional frost, but there is little if any snow. In some years there is so little rain that jack rabbits wander across the dry beds of good-sized rivers.

The most characteristic plant of the Sonoran Province is the creosote bush (*Larrea divaricata*, Fig. 22.88), a branching, aromatic shrub commonly 3 or 4 feet tall with a very deep root system and with small, shiny, firm, evergreen leaves. Much of the northern boundary of the Sonoran Province is sharply marked by the limit of the creosote bush community. In some places, especially in Arizona, the boundary coincides with an abrupt drop of more than a thousand feet from the Great Basin to the Sonoran Province, and the transition from one province to the other takes only a mile or so.

Over thousands of square miles of this province, in both the United States and Mexico, creosote bushes are spaced at more or less regular intervals of 15 to 30 feet, with much bare ground in between. Bur sage (*Franseria dumosa*), another low, branching shrub, is often associated with creosote bush, and cacti of various sorts are common.

Ephemeral annuals, many with showy flowers, are abundant in the warm desert. In a good year they color the landscape purple, yellow, and white for a time before the plants mature seed and disappear. If the rains fail, the plants may not come up at all, or they may wither and die before flowering. The seeds of some species will not germinate unless there is enough moisture in the soil to last through a normal cycle of growth and fruiting.

The seeds of most desert annuals will not all germinate the first season

Fig. 22.88. Creosote bush. This same species grows in deserts south of the equator in South America as well as in our Sonoran Province. (Photo by Allan B. Lang, from National Audubon Society)

Fig. 22.89. Creosote bush desert community in the Mojave desert near Lathrop Wells, Nevada. Note the expanses of bare ground between the bushes. Some of the smaller bushes at the left are bur sage.

after they are formed, no matter how favorable the conditions. Some will germinate the first year, but others of the same lot require one, two, three, or more years of dormancy before they will grow. Otherwise the whole population would be subject to being wiped out during some year in which normal conditions at the beginning of the spring are followed by a catastrophic drought.

Different kinds of desert plants show diverse adaptations to conditions of water shortage. In general, they may be classified into three groups: (1) ephemerals, (2) succulents, and (3) drought-enduring species.

Ephemerals are short-lived annuals which complete their life cycle during a few weeks or months when the soil is moist (typically in the spring) and survive the dry period as seeds. During the period of their activity, many ephemerals are no more drought-resistant than plants of moister regions; they are drought-escaping rather than drought-enduring. Ephemerals, like other annuals, have the competitive disadvantage, as compared to perennials, of having to produce an average of one mature descendant every year, instead of one every several to many years. In areas where the ground is more or less covered with vegetation throughout the year, annuals have a hard time. In desert regions, however, and especially in warm deserts such as those of the Sonoran Province, there is a good deal of bare ground, and many of the seeds therefore do not have to compete directly against plants which are already established.

Succulents are characterized by the accumulation of reserves of water in the fleshy stems or leaves, due largely to the presence of a high proportion of water-retaining colloids in the protoplasm and cell sap. Some of them, such as the cacti, also have very much reduced leaves, and the photosynthetic parts generally have a very thick cuticle which reduces evaporation of water.

Drought-enduring species are characterized by the ability to endure desiccation without irreparable injury. Creosote bush is a good example of a drought-enduring species. During dry periods the water content of the leaves may sink to less than 50 percent of their dry weight, in contrast to the leaves of most plants of moister regions, in which the water content generally ranges from 100 to 300 percent of the dry weight. Under such conditions the leaves are of course largely dormant; only when the water supply again becomes adequate do they return to physiological activity.

Many drought-enduring species have firm leaves with a high proportion of strengthening tissue. Such plants, as we have pointed out in an earlier chapter, are called sclerophylls. Experimental studies have shown that under

Fig. 22.90. Joshua trees at the edge of the Mojave Desert in California. (U.S. Forest Service photo by A. E. Wieslander)

gusty wind conditions water evaporates less rapidly from such leaves than from more flexible ones, but the importance of the difference under natural conditions is doubtful. It is perhaps more important that the strengthening tissue serves to prevent mechanical injury when the leaf is physiologically wilted during periods of drought.

A variety of other adaptations aid drought resistance in some species. Frequently the aerial part of the plant is very small in proportion to the root system. The cuticle is commonly rather thick, and the stomates (through which gases are exchanged between the internal tissues of the leaf and the outer atmosphere) are sometimes sunk in pockets or protected by a matting of hairs. The leaves sometimes fall off during the dry season, minimizing the loss of water during the period of shortest supply.

Desert plants differ among themselves in the part of the soil which they exploit for water. Many of the shrubs, such as sagebrush and creosote bush, are notoriously deep-rooted. If the soil has permanent moisture at any level, they will find it. Cacti, on the other hand, commonly have most of their roots in the upper few inches of soil, and these may spread for many feet in all directions from the base of the plant. Thus they are able to take advantage of any light rain which may fall, even though the water does not penetrate deeply into the soil.

The Sonoran Province as we have defined it could be divided into several closely related provinces, all of them with scattered mountains arising from lowland deserts. The northwestern part of the province, in California, Nevada, and northwestern Arizona, is the Mojave Desert. Farther south we have, from west to east, the Sonoran, Chihuahuan, and Tamaulipan Deserts. Each of these last three deserts reaches its northern limits in our border states and extends well down into Mexico.

The unifying factor for all these subprovinces except the Tamaulipan is the creosote bush, which occurs in all the lowland areas. The Tamaulipan Desert, which enters the United States only at the southern tip of Texas, is not quite so dry as the others and is characterized by a thorny scrub growth in which mesquite (*Prosopis*) is a prominent shrub or small tree. Mesquite also extends westward into the Chihuahuan Desert in decreasing abundance, but it is not significant in the Sonoran desert and does not reach the Mojave Desert at all.

In the Mojave and Sonoran Deserts the rainfall, such as it is, comes mainly in the winter and early spring. The big burst of flowering is in the spring, when the weather is good and the ground is still moist. In a good year the hills around Hoover Dam on the Colorado River are a mass of mixed colors,

Fig. 22.91. Saguaro cactus in southern Arizona. (New York Botanical Garden photo)

Fig. 22.92. Paloverde (*Cercidium*) in the desert in Arizona. The twigs are bright green even during the dry season when the leaves have fallen. Some species have somewhat yellowish-green bark, and others are more bluish-green. (U.S. Forest Service photo by Rex King)

and the same sort of display can be seen in much of the western part of the whole Sonoran Province. A large proportion of the flowers are annuals, and many of these are actually winter annuals, which germinate after the first significant rains of the winter and burst into bloom when the weather warms up in the spring.

Farther east, in the Chihuahuan and Tamaulipan deserts, the rainfall comes mainly in late summer or fall, and the big burst of bloom is in the fall. Tucson, Arizona, is near the dividing line and gets some rain both summer and winter, though not enough to support anything other than desert vegetation. A person familiar with the deserts of southern California would feel reasonably at home in Tucson in the spring, but he would be lost in late summer and fall, when the species of Chihuahuan affinities begin to bloom.

Some of the most interesting plants of the Sonoran Floristic Province are giant Joshua (*Yucca brevifolia*), the saguaro cactus (*Carnegiea gigantea*), the ocotillo (*Fouquieria splendens*), the paloverde (*Cercidium*), and the century plant (*Agave americana*). None of these occurs throughout the whole area, but each is a conspicuous member of the flora in some part of the province.

The giant Joshua (Fig. 22.90) is a grotesque shrub or small tree, which one disillusioned explorer said casts about as much shade as a barbed wire fence. It occurs in the northern and western parts of the Sonoran Province, especially in the Mojave Desert region. It characteristically grows at the upper margin of the creosote bush community, often forming a fringe only a mile or so wide where the cool desert of the Great Basin Province slopes down to the warm desert of the Sonoran Province. In southeastern California an area dominated by giant Joshua has been set aside as a national monument. Several other species of yucca which resemble the giant Joshua but tend to be smaller and less branched grow in the creosote bush community in parts of Mexico.

Yuccas in general are pollinated by moths belonging to the genus *Pronuba*. The female moth gathers the pollen, rolls it into a little ball, and pats it onto the stigma of the flower, giving every outward evidence of knowing exactly what it is doing. It then lays eggs in the ovary of the same or another flower. The moth larvae eat some of the developing seeds within the ovary, but others ripen undisturbed. If you are tempted to cut off a yucca inflorescence for a bouquet, be prepared to see little green worms crawling over it a day or so later when the changed conditions impel them to wander in search of a better home.

The saguaro (Fig. 22.91) is the largest of all cacti, reaching a height of about

Fig. 22.93. Palms (*Washingtonia filifera*) near Palm Springs in southern California. This species, more frost-resistant than many palms, is often cultivated in the warmer parts of southwestern United States. (National Park Service photo)

Fig. 22.94. Semidesert grassland in the Santa Rita Mountains in southeastern Arizona. (U.S. Forest Service photo by E. C. Crafts)

50 feet. It is abundant in parts of southern Arizona and northern Sonora (Mexico), where it is generally the only arborescent plant of any sort and is commonly inhabited by woodpeckers. Other cacti nearly as large as the saguaro occur in the more southern parts of the warm desert.

Ocotillo (Fig. 18.3) is a picturesque tall shrub with several arching, wand-like branches from the base. Its numerous rather delicate leaves are produced whenever the soil is moistened by a substantial rain, and they fall off again when the soil dries out. The flame-red flower-clusters terminate the branches in March and April.

Species of the genus *Cercidium* are commonly known as paloverde (Fig. 22.92). They are robust, spiny shrubs or small trees with conspicuously green twigs (whence the common name). They are leafless for most of the year, and the leaves when present are relatively small and insignificant. The plants have yellow flowers and are very showy when in bloom. Various species of paloverde are common in the Tamaulipan, Chihuahuan, and Sonoran Deserts, but they scarcely reach the Mojave.

The century plant (Fig. 5.2) flowers only once during its lifetime. During the period of its vegetative growth it has a cluster of large leaves at the ground level, but no obvious stem. Then, after 10 to 25 or more years (but not a century, as the name implies), it rapidly sends up a tall stem with a large, terminal inflorescence. The plant dies as the seeds mature.

Several other species of *Agave* and some species of Yucca have the same general aspect as the century plant and behave in much the same way. Some years ago a national scientific magazine published a mock-serious article about a kind of Yucca which grew so fast that an animal which jumped over the plant at the right (or wrong) time would be impaled on the growing stem. Some readers recognized the joke and wrote in with further elaborations in the style of the Baron Munchausen, but others were less perceptive and wrote ponderous rebuttals. The editor lost his job.

The mountains and other highlands of the Sonoran Province are, of course, cooler and moister than the intervening lowlands, and the difference in climate is, as always, reflected in the vegetation. In the more eastern part of the province, the next zone above the true desert is often semidesert grassland (Fig. 22.94). Toward the west it is more often oak brush or chaparral resembling the chaparral of the Californian Province. Above this there may be a zone of junipers and piñon pines, which in turn gives way to a more characteristic coniferous forest. The highest peaks extend above timber line and have a more or less alpine vegetation. Toward the northern part of the province the

Fig. 22.95. Two common grasses in the semidesert ranges of southwestern United States. Left, slender grama (*Bouteloua filiformis*); right, galleta (*Hilaria jamesii*). (U.S. Soil Conservation Service photos by Hermann Postlethwaite)

species in the upper vegetational zones are largely the same as those of comparable habitats in the Rocky Mountains and Great Basin; there are, for example, many Rocky Mountain species in the mountains of trans-Pecos Texas. Southward, these are progressively replaced by other species, and the traveler from the United States will find fewer and fewer familiar plants as he goes on into Mexico.

Index

NOTE: *In keeping with the practice in the foregoing text, plants are here indexed under their English names so far as is reasonably possible, and the scientific names are appended for accurate reference. When a well-established English name exists, the scientific name is usually not separately indexed.*

Abundance, concept of, 78
Acanthella conferta, 20
Acer torontoniensis, 115
Aggressive species, 75 f.
Albino flowers, 115
Alder (*Alnus*), 279
Alkali, 375
Altitudinal zonation, 9, *10*, 11, 353
Anemone (*Anemone*), 307; rue (*Anemonella thalictroides*), 162 ff.; wood (*Anemone quinquefolia*), 48 f.
Annuals, 61, 226, 400 f.
Appalachian region, 317 f.
Aquatic plants, 139 ff., *142*
Arctic Province, *see* Tundra Province
Ash (*Fraxinus*), 28, 333; black (*Fraxinus nigra*), *29;* white (*Fraxinus americana*), 154, 160
Aspen, quaking (*Populus tremuloides*), *21*, 134, *135*, 293, *295*, 352, 353, 372, *386*
Aspen-grass woodland, *351*, 352
Asphodel, bog (*Narthecium americanum*), 78
Aster (*Aster*), 307, 336, 346, 372; New England (*Aster novae-angliae*), 23, *24*
Aster family (Compositae), 307, 336, 346, 398
Autochthons, 179 ff.

Bald cypress (*Taxodium*), *17*, 102, *106*, 186, *327*, 333
Balds, 244, *301*
Baldwin, William, 82
Balsam root, arrow-leaved (*Balsamorhiza sagittata*), *378*, 387
Basswood (*Tilia*), 312
Bay, red (*Persea Borbonia*), 341; sweet (*Magnolia virginiana*), 330, 341
Beardtongue (*Penstemon*), 372, 387
Beech, American (*Fagus grandifolia*), 91, *93*, 96, 154, 159, 312, 330
Beet family (Chenopodiaceae), 384
Bessey, C. E., 92
Biennials, 61, 64
Bigtree, *see* Sequoia, giant
Birch (*Betula*), 279, 353; gray (*Betula populifolia*), *132*, 134; paper (*Betula papyrifera*), 192, *195*, 293, *294*; river (*Betula nigra*), 316
Black-eyed Susan (*Rudbeckia hirta*), *117*, 119
Black Hills, 352 f.
Blanket-flower (*Gaillardia*), 346
Bloodroot (*Sanguinaria canadensis*), 136, *141*, 162, 164 f.
Bluegrass (*Poa*), 280
Bluestem, big (*Andropogon Gerardi*), 55, 346, *349;* little (*Andropogon scoparius*), *217*, 222, 346
Bracken fern, *see* Fern, bracken
Bromeliaceae, *342*
Buckeye (*Aesculus glabra*), 186, *190*, *312*
Buffalo bur (*Solanum rostratum*), 33
Bur sage (*Franseria dumosa*), 400
Buttercup (*Ranunculus*), 280, 372
Buttercup family (Ranunculaceae), 297
Butterweed (*Senecio*), 346

Cactus, 86, 405; giant, or saguaro (*Carnegiea gigantea*), *178*, 179, *406*, 408 f.
Calciphiles, 96
Calciphobes, 96
California laurel, *see* Laurel, California
California lilac (*Ceanothus*), 391, *393*
Californian Province, 176, 388 ff.
Calypso bulbosa, 293, *296*
Campion (*Silene*), 280
Cardinal flower (*Lobelia Cardinalis*), 118
Carex, 280
Caryophyllaceae, 297
Catalpa (*Catalpa*), 28, *30*

Cattail (*Typha*), 39, *143*
Cedar, eastern red (*Juniperus virginiana*), 316; northern white (*Thuja occidentalis*), 316, *317;* southern white (*Chamaecyparis thyoides*), *331*, 333; western red (*Thuja plicata*), 355, *358*
Cedar glades, 316
Century plant (*Agave americana*), 61, *63*, 226, 411
Chamaephytes, 299
Chamise (*Adenostoma fasciculatum*), 391
Chaparral, *390*, 391, 398, 411
Chaparral Province, *see* Californian Province
Chenopodiaceae, 384
Chestnut, American (*Castanea dentata*), 88, *89*, 90, 312 f.
Chickweed (*Stellaria media*), 75
Chinquapin, giant (*Castanopsis chrysophylla*), 392
Cinnamon, wild (*Canella Winteriana*), 342
Classification, 129 ff.
Climate, 239 ff.; desert, 254 ff.; forest, 240 ff.; grassland, 251 ff.; Mediterranean, 258 ff.; sclerophyllous forest, 257 ff.; tundra, 253 ff.
Climatic changes (*table*), 250–51
Climax community, 275 f.
Coastal Plain Province, 176, 216 ff., 321 ff.
Coconut (*Cocos nucifera*), 31, *35*
Columbine (*Aquilegia*), *389*
Commonness, concept of, 78
Communities, plant, 133 ff.; *see also* Forest, Desert, etc.
Compass plant (*Silphium laciniatum*), 197, *198*
Compositae, 307, 336, 346, 398
Coneflower (*Rudbeckia* and *Ratibida*), 346
Cordilleran Forest Province, 176, 353 ff.
Cornel, dwarf (*Cornus canadensis*), 297, *299*
Cotton grass (*Eriophorum*), 280, *283*
Cottonwood (*Populus deltoides*), *262*, 316
Creosote bush (*Larrea divaricata*), 400, *401*, *402*
Crowberry, broom (*Corema Conradii*), 78
Cryptophytes, 229
Curly grass (*Schizaea pusilla*), 221
Curtis, John T., 272
Custard apple (*Annona glabra*), 336
Cypress, bald, *see* Bald cypress; Monterey (*Cupressus macrocarpa*), 392, *395*

Daisy (*Erigeron*), 280, 372; oxeye (*Chrysanthemum Leucanthemum*), 61
Dandelion (*Taraxacum*), 28
Decadent species, 75 f.
Desert, 225 f., 254 ff.; Chihuahuan, 405 f.; cool, 375 ff.; Mojave, 405 f.; Sonoran, 405 ff.; Tamaulipan, 405 f.; warm, 400 ff.
Devil's club (*Oplopanax horridum*), 357
Dilly, wild (*Manilkara emarginata*), 342
Dispersal mechanisms, 27 ff.
Dodds, G. S., 91, 201
Dogtooth violet (*Erythronium*), 162 ff., *166*
Douglass, A. E., 72
Drought-endurance, 403
Duckweed (*Lemna* and *Spirodela*), 15
Dutchman's-breeches (*Dicentra Cucullaria*), 162 ff., *163*

Eastern Deciduous Forest Province, 176, 302 ff.
Elm, American (*Ulmus americana*), *19*, *45*, 47, 128, 146, *147*, 154, 160, 316
Endemism, 180 f.
Ephemerals, 400, 403
Epiphytes, 342, *343*
Ericaceae, 279, 293, 372
Eucalyptus (*Eucalyptus*), 240, *259*
Evening primrose family (Onagraceae), 398
Everglades, 341 f.

Fern, bracken (*Pteridium aquilinum*), *22*, 23; maidenhair (*Adiantum pedatum*), 134, *138*
Fescue, sheep (*Festuca ovina*), 384
Fig (*Ficus*), 342
Fir, alpine (*Abies lasiocarpa*), 364, *366;* balsam (*Abies balsamea*), 287, *289;* Douglas (*Pseudotsuga Menziesii*), 355, *356*, 364, *367*, *368;* Fraser (*Abies Fraseri*), 297; red (*Abies magnifica*), 357, *361*
Fire, 104, 206, 287, 316, 321 ff., 345, 355, 372, 391
Fireweed (*Epilobium angustifolium*), *46*, 48
Fireweed (*Erechtites hieracifolia*), 37
Fleabane, marsh (*Pluchea*), 37
Flood plains, 316, *319*
Flora, 175 ff.
Floristic group, 176
Floristic Province, 175 ff.; map, *174*
Flowering rush (*Butomus umbellatus*), 84
Fog belt, 264 f.
Forest, 235 f., 240 ff.; beech-maple, 312; beech-maple-hemlock, 74, *314*, 316; birch-maple-basswood, *310;* boreal, 285 ff.; broad sclerophyll, 392 f.; closed-cone pine, 392; cove, *308*; deciduous, 302 ff.; Douglas fir, 355, *356;* gallery, *262*, 264; lodgepole pine, *369;* maple-basswood, 312; mixed hardwood, 330, *344;* mixed mesophytic, *308;* mossy, 244, *245;* oak-chestnut, 312 f.; oak-hickory, *311*, 312; pine, *320*, 321 ff.; ponderosa pine, *363;* rain, 353; red fir, *361;* redwood, *241*, 355; sclerophyllous, 257 ff., 392 f.; spruce-fir, 287, *290*, 297; spruce-hemlock, 357, *360;* western red cedar, *358;* western white pine, *359*, white pine, *315*, 316
Forms, life, 229 ff.; vegetative, 228 ff.
Franklin tree (*Franklinia alatamaha*), 20
Frost line, 102

Galax aphylla, 120
Galinsoga ciliata, 82 f.
Galleta (*Hilaria Jamesii*), *413*
Geographic races, 115 f.
Geranium, wild (*Geranium maculatum*), 162 ff.
Giant Joshua, *see* Joshua tree
Giant sequoia, *see* Sequoia, giant
Ginger, wild (*Asarum canadense*), 162 ff., *172*
Ginseng (*Panax quinquefolium*), *304*
Goldenrod (*Solidago*), *112*, 116, 307, 312, 336, 346
Grass, beach (*Ammophila arenaria*), *234*, 236; blue, *see* Bluegrass; blue grama, *see* little grama; buffalo (*Buchloe dactyloides*), *219*, 222, 346; cord (*Spartina*), 118; cotton, *see* Cotton grass; curly, *see* Curly grass; Indian (*Sorghastrum nutans*), 346; June (*Bromus tectorum*), 384; little grama (*Bouteloua gracilis*), *220*, 346; saw, *see* Saw grass; slender grama (*Bouteloua filiformis*), *412*; wheat, *see* Wheatgrass; wool, *see* Wool-grass
Grassland, *249*, 251 ff., 353, 364; Californian, *396*, 397; semidesert, *410*, 411
Grassland Province, 176, 221 f., 345 ff.
Greasewood (*Sarcobatus vermiculatus*), 375
Great Basin Province, 176, 372 ff.
Great Plains, 348
Greenbrier (*Smilax*), *334*, 336
Griggs, R. F., 91
Gum (*Nyssa*), *328*, 333; sweet, *see* Sweet gum
Gumbo limbo (*Bursera Simaruba*), *340*, 341 f.

Habit, 226 ff.
Habitat, 56
Halogeton, *377*
Hammocks, 341 f.
Hawkweed (*Hieracium*), 61
Heather, white (*Cassiope Mertensiana*), *278*
Heath family (Ericaceae), 279, 293, 372
Hemicryptophytes, 229
Hemlock, eastern (*Tsuga canadensis*), 74 f., 182, *187*; mountain (*Tsuga Mertensiana*), 355; western (*Tsuga heterophylla*), 355
Hemp nettle (*Galeopsis tetrahit*), 118
Hepatica (*Hepatica americana*), 162 ff., *169*
Hickory (*Carya*), 312, 330; bitternut (*Carya cordiformis*), 154, 159; shagbark (*Carya ovata*), 154, *156*, 159
Holly (*Ilex opaca*), 330
Hooker, Sir Joseph Dalton, 5
Humboldt, Baron Alexander von, 9 f.
Hyacinth, water, *see* Water hyacinth
Hybridization, 114, 118 ff.
Hydrophyllaceae, 398

Iliamna remota, 20, 56, 58
Indian paintbrush (*Castilleja*), *371*, 372, 387
Indian pipe (*Monotropa uniflora*), 37, *298*
Indian turnip, *see* Jack-in-the-pulpit
Ironweed (*Vernonia*), 221

Jack-in-the-pulpit (*Arisaema triphyllum*), 162 ff., *170*
Jacob's ladder (*Polemonium*), 280
Jewel weed, *see* Touch-me-not
Joe-pye weed (*Eupatorium maculatum*), *204*, 205
Joewood (*Jacquinia keyensis*), 342
Jordan, David Starr, 123
Joshua tree (*Yucca brevifolia*), *255*, *404*, 408
Juniper, Rocky Mountain (*Juniperus scopulorum*), 387; Utah (*Juniperus osteosperma*), *380*, 387

King devil (*Hieracium aurantiacum*), 33
Klamath region, 398
Krummholz, *101*

Lady slipper, showy (*Cypripedium reginae*), 76, *77*
Lakes, transitory nature, 136 ff.
Land of little sticks, 297
Larch, *see* Tamarack
Larkspur (*Delphinium*), 372
Laterite, 302
Laurel, California (*Umbellularia californica*), 392 f.; mountain, *see* Mountain laurel
Leatherleaf (*Chamaedaphne calyculata*), *54*, 55, 78
Leguminosae, 346
Lichens, 279
Life forms, 229 ff.
Lily, lotus, *see* Lotus lily; sego (*Calochortus Nuttallii*), *379*, 387
Lily family (Liliaceae), 398
Live oak, *see* Oak, live
Lobelia, great blue (*Lobelia siphilitica*), 37, 118
Locoweed (*Astragalus*), 387
Loosestrife, purple (*Lythrum Salicaria*), *83*, 84
Lotus lily (*Nelumbo lutea*), *142*
Lousewort (*Pedicularis*), 280
Lupine (*Lupinus*), *370*, 372, 387

Madrone (*Arbutus Menziesii*), 357, 392 f.
Magnolia, 312; *see also* Bay, sweet, and Umbrella tree
Mahogany (*Swietenia Mahogani*), 342; mountain, *see* Mountain mahogany
Maidenhair fern, *see* Fern, maidenhair
Mallee, 260
Mangrove, 336, *337*, 338 ff.; black (*Avicennia nitida*), 341; red (*Rhizophora Mangle*), 338 f.
Manzanita (*Arctostaphylos*), 391
Maple (*Acer*), 312, 353; silver (*Acer saccharinum*), 39, 146, *148–49*, 154, 160, 316; sugar (*Acer saccharum*), 15, *18*, *66*, 154, *157*, 160, *309*, 312
Marsh fleabane, *see* Fleabane, marsh

Mastic (*Mastichodendron foetidissimum*), 341, 342
May-apple (*Podophyllum peltatum*), 68 f., *69*, 162 ff.
Meadow, 235 f.
Meadow rue (*Thalictrum dioicum*), 162 ff.
Mesquite (*Prosopis*), *85*, 86, 104 f., *106–107*, 405
Metasequoia, 76, *79*
Migration, 27 ff.; speed of, 81 ff.
Milkweed (*Asclepias*), *26*, 28
Mistletoe (*Phoradendron*), 99, *103*
Monkey flower (*Mimulus*), 37
Moss, 279; hairy-cap (*Polytrichum*), 280; peat (*Sphagnum*), 280, 333; reindeer, *see* Reindeer moss; Spanish, *see* Spanish moss
Mountain laurel (*Kalmia latifolia*), *189*, *190*
Mountain mahogany (*Cercocarpus ledifolius*), *385*, 388
Mullein (*Verbascum Thapsus*), *60*, 61
Muskeg, *144*, 285, *286*
Mutation, 114 ff.
Mycorhiza, 293
Myrtle, Oregon, *see* Laurel, California

Nettle, hemp, *see* Hemp nettle
Neviusia alabamensis, 20
Northern Conifer Province, 176, 285 ff.

Oak (*Quercus*), 42, 312, 330, 353; black (*Quercus velutina*), 154, 159; bur (*Quercus macrocarpa*), *347*, 352; California black (*Quercus Kelloggii*), 392; canyon (*Quercus chrysolepis*), 391 f.; interior live (*Quercus Wislizenii*), 391 f.; live (*Quercus virginiana* and other species), *7*, *213*, 218, *324*, 330, 391 f.; northern red (*Quercus borealis*), 154, *155*, 159; Oregon (*Quercus Garryana*), 392; scrub (several species of *Quercus*), *326*, 333, 388, 391; shingle (*Quercus imbricaria*), *207*, 208; swamp white (*Quercus bicolor*), 182, *184*, *185*; tanbark (*Lithocarpus densiflora*), 392; valley (*Quercus lobata*), 392; white (*Quercus alba*), 154, 159, *224*
Oak-hickory forest, *see* Forest, oak-hickory
Oak woodland, 391 f., *394*, 398, *399*, 411
Oats, wild (*Avena barbata*), 34, *396*, 397
Ocotillo (*Fouquieria*), *231*, 408, 411
Onagraceae, 398
Orchid family (Orchidaceae), 293 f., 342, 372
Oregon myrtle, *see* Laurel, California
Origin of species, 113 ff.
Osage orange (*Maclura pomifera*), *43*, 44
Ozark region, 318

Paintbrush, Indian, *see* Indian paintbrush
Palm, *409*; *see also* Palmetto
Palmetto, cabbage (*Sabal Palmetto*), *332*, 333; dwarf (*Sabal minor*), *215*, 333
Palouse region, 364, *365*
Paloverde (*Cercidium*), *407*, 411
Paramo, 285
p/e ratio, 387
Pea family (Leguminosae), 346
Peat bogs, 205, 214
Pecan (*Carya illinoensis*), 186, *193*
Pennyroyal, western (*Hedeoma hispida*), 62
Perennials, 61 f.
Permafrost, 253 f., 276
Phacelia, 387
Phanerophytes, 229
Phlox (*Phlox*), 123, 387
Phlox family (Polemoniaceae), 398
Pine, bishop (*Pinus muricata*), 392; Coulter (*Pinus Coulteri*), 392; digger (*Pinus Sabiniana*), 392; jack (*Pinus Banksiana*), 287, *292* Jeffrey (*Pinus Jeffreyi*), 357; limber (*Pinus flexilis*), *101*, 364; loblolly (*Pinus taeda*), *325*; lodgepole (*Pinus contorta*), 364, *369*, 372; longleaf (*Pinus australis*), 186, *191*, *320*, 321; Monterey (*Pinus radiata*), 392; nut, or piñon (*Pinus edulis* and *Pinus monophylla*), *381*, 387 f.; pitch (*Pinus rigida*), 134, *137*; pond (*Pinus serotina*), *329*; ponderosa (*Pinus ponderosa*), 91 f., *94*, *200*, 357, *363*, 364; red (*Pinus resinosa*), 54; single-leaf piñon (*Pinus monophylla*), *381*, 387 f.; slash (*Pinus caribaea*), 338, 345; sugar (*Pinus Lambertiana*), 357, *362*; Torrey (*Pinus Torreyana*), 20, 186, *194*; two-leaved piñon (*Pinus edulis*) 387 f.; western white (*Pinus monticola*), *73*, 355, *359*; western yellow, *see* ponderosa; white (*Pinus Strobus*), 55, *57*, *210*, 212, *315*, 316; whitebark (*Pinus albicaulis*), 364, *366*
Pine barrens, 322 ff., *323*
Pine forest, *see* Forest, pine
Pink family (Caryophyllaceae), 297
Piñon, *see* Pine, piñon
Piñon-juniper woodland, 353, 364, *382*, *383*, 387
Pinweed (*Lechea*), 71
Pitcher plant (*Sarracenia*), 14, 78, *121*, *122*, 125 f., 221; California (*Darlingtonia californica*), *14*
Plant communities, 133 ff.
Plant succession, 136 ff.
Playa lakes, 372
Pocosin, *329*, 333
Podzol, 285, 302
Polemoniaceae, 398
Pollen, fossil, 214
Ponds, transitory nature of, 136 ff.
Poplar (*Populus*), balsam (*Populus balsamifera*), 293; yellow, *see* Tulip tree; *see also* Cottonwood
Poppy, California (*Eschscholzia californica*), 398
Potato, sweet (*Ipomoea batatas*), 33, *36*
Prairie, mixed-grass, 346, 348, *350*; short-grass, *227*, 346, 348; tall-grass, 346

Prairie peninsula, 345
Prairies and Plaines Province, *see* Grassland Province
Puerto Rico, vegetation in, 272 f.

Races, geographic, 115 f.
Ragweed (*Ambrosia*), *306*, 307 f.
Ramaley, Francis, 91, 201
Range, joint, 154
Ranunculaceae, 297
Rarity, concept of, 78
Raspberry, wild (*Rubus*), 48
Raunkiaer, Christen, 228 ff.
Redwood (*Sequoia sempervirens*), 186, 239, *241*, 355; Sierra, *see* Sequoia, giant
Reed (*Phragmites communis*), 39, *41*
Reindeer moss (*Cladonia*), 280, *282*
Relic colonies, 211 f.
Rhododendron catawbiense, *301*
Rose family (Rosaceae), 297
Rue, meadow, *see* Meadow rue
Rue anemone, *see* Anemone, rue
Rush, flowering, *see* Flowering rush

Sage, bur, *see* Bur sage
Sagebrush (*Artemisia tridentata*), 202, *203*, 353, *373*, *374*, 375, 384
Sagebrush-grass community, *373*, *374*
Saguaro, *see* Cactus, saguaro
Sapotaceae, 342
Satinleaf (*Chrysophyllum olivaeforme*), 342
Savanna, 312
Saw grass (*Mariscus*), 338, *339*, 341
Saxifrage (*Saxifraga*), 280
Saxifrage family (Saxifragaceae), 297
Schneck, Jacob, 100
Sclerophylls, broad, 391 ff.
Seed dispersal, 27 ff.
Sequoia, giant (*Sequoiadendron giganteum*), 6, *8*, 357, 364
Serpentine, 397 f.
Shadscale (*Atriplex*), 375, *376*
Shooting star (*Dodecatheon*), 372
Sidesaddle flower (*Sarracenia purpurea*), 125; *see also* Pitcher plant
Silver-bell tree (*Halesia carolina*), 182, *183*
Silverweed (*Potentilla anserina*), 52
Skunk cabbage (*Symplocarpus foetidus*), 37, 52, *53*
Snake River plains, *373*
Snakeroot, white (*Eupatorium rugosum*), *305*
Sonoran Province; 177; 400 ff.
Sotol (*Dasylirion*), *231*
Spanish moss (*Tillandsia usneoides*), 6, *7*, *17*, 100 f., 333 f., 342
Sparsity, concept of, 78
Species, origin of, 113 ff.
Spectrum, normal, 230
Sphagnum, *see* Moss, peat
Sports, 114
Spring beauty (*Claytonia*), 162 ff., *303*, 307
Spruce, black (*Picea mariana*), *50*, 51, 151, *152*, 287, *291*; Colorado blue (*Picea pungens*), 364; Engelmann (*Picea Engelmannii*), 101, *243*, 364, *366*; red (*Picea rubens*), 297, *300*; Sitka (*Picea sitchensis*), 357; white (*Picea glauca*), 287, *288*
Steere, W. C., 254
Succession, plant, 136 ff.
Succulents, 403
Sumac, common (*Rhus glabra*), 208, *209*
Sunflower (*Helianthus*), 346
Sweet gum (*Liquidambar styraciflua*), 316
Sweet potato, *see* Potato, sweet
Sweet William (*Phlox divaricata*), 162 ff.
Sycamore (*Platanus occidentalis*), 96, *97*, *98*, 316

Tamarack (*Larix laricina*), 151, *153*, 287
Tanbark oak, *see* Oak, tanbark
Territoriality, 16
Therophytes, 229
Tillandsia, *see* Spanish moss
Timber line, *see* Tree line
Toothwort (*Dentaria laciniata*), 162 ff.
Touch-me-not (*Impatiens*), 34 f., 39, *40*
Toumey, J. W., 67
Transitions between provinces, 263 ff.
Tree line, 99, *101*, 242, *243*, 244 ff., 285, 297, 357, 364, 388
Tree of heaven (*Ailanthus altissima*) 34, *38*
Trillium (*Trillium*), 162 ff., *171*; dwarf white (*Trillium nivale*), 123; painted (*Trillium undulatum*), 123
Tulip tree (*Liriodendron Tulipifera*), 57 f., *59*, 154, *158*, 160, 182, 192, *196*, 312
Tumbleweed (*Amaranthus albus*, *Salsola kali*, and others), 28, *32*, 62 f.
Tundra, 253 ff., 276 ff., *277*; alpine, 280, *284*, 285, 372, 388
Tundra Province, 176, 276 ff.

Umbrella tree (*Magnolia macrophylla*), *181*, 182

Vegetation, 235 ff.
Venus's-flytrap (*Dionaea muscipula*), 221, *335*, 336
Violet (*Viola*), 307; dogtooth, *see* Dogtooth violet; yellow (*Viola pubescens*), 162 ff.

Walnut, black (*Juglans nigra*), 154, 160
Water hyacinth (*Eichhornia crassipes*), 15, *17*
Waterleaf (*Hydrophyllum virginianum*), 162 ff.
Waterleaf family (*Hydrophyllaceae*), 398
Water lettuce (*Pistia*), 15

Wax myrtle (*Myrica cerifera*), 341
West Indian Province, 176, 336 ff.
Wheatgrass, bluebunch (*Agropyron spicatum*), *354*, *374*, 384
Willow (*Salix*), 279, 280; black (*Salix nigra*), 316
Winter fat (*Eurotia lanata*), *376*, 384
Wintergreen (*Pyrola*), 293
Wiregrass (*Aristida*), 321
Wool-grass (*Scirpus* species), *331*

Xerothermic period, 206

Yucca (*Yucca*), 411; *see also* Joshua tree

Zonation, altitudinal, 9, *10*, 11, 353